Nigel Dudley

The Death of Trees

Pluto Press

London and Sydney

First published in 1985 by Pluto Press Limited,
The Works, 105a Torriano Avenue, London NW5 2RX
and Pluto Press Australia Limited, PO Box 199, Leichhardt,
New South Wales 2040, Australia

Cover designed by Clive Challis A.Gr.R

Phototypeset by A.K.M. Associates (U.K.) Ltd,
Ajmal House, Hayes Road, Southall

Printed in Great Britain by
Guernsey Press, Guernsey C.I.

British Library Cataloguing in Publication Data
Dudley, Nigel
    The death of trees.
    1. Clear-cutting
    I. Title
    333.75   SD538

ISBN 0 86104 613 7

Contents

Preface

This book is about forests and politics. Specifically, it is about how forests throughout the world are being destroyed at an accelerating rate and what social and political factors are combining to cause this destruction. The text is the result of discussion over a number of years within various organizations, including Friends of the Earth, World Forest Action and Earth Resources Research. It aims to be a fairly concise introduction to the subject that will appeal both to environmentalists without much knowledge of politics, and people who are interested in agriculture and development but who know little about nature conservation.

Until recently, environmentalists have tended to steer clear of the political causes underlying pollution and habitat destruction. Traditionally, nature conservationists have preferred to carry out practical conservation in the country than to fight political battles in the city. This has enormously weakened the conservation case. There is now a groundswell within the green movement which is beginning to look beyond the depressing headlines about the environment, to grasp the problem of eco-catastrophe at its roots, in the powerful capitals of the developed world. It is seldom the poor and the powerless in the Third World (or anywhere else) who decide the fate of forests or other natural resources. Irreplaceable ecosystems are cynically manipulated and destroyed for the profits of a minority, both by direct exploitation of trees and by using forest areas as a dumping ground for unwanted peoples who have been expelled from better land elsewhere. It is hoped that this book can be a voice in the growing debate about control and management of the planet.

Acknowledgements

Many people have generously given advice at all stages of the book. I would like to thank in particular Herbert Giradet, Robert Lamb and Sally Boyle of World Forest Action, Charles Secrett of Friends of the Earth, François Nectoux and David Baldock of Earth Resources Research and Pete Ayrton at Pluto Press for helping to shape my research into publishable form. I am especially indebted to the writings of Norman Myers and Catherine Caufield, many of whose examples have been incorporated into the following account.

Introduction

Supposing that you spend a leisurely three hours reading this book; in the time it takes you to do so, some 5,500 hectares of unique tropical forest will have been destroyed somewhere in the world. Statistics like this are gross simplifications of course, but as near as we can judge, an area about the size of England and Wales is felled every year; put another way, this is about a football pitch's worth of forest per second, and the rate is still increasing. To date, about 40 per cent of primary (i.e. untouched) tropical forests have already been destroyed by machetes, chainsaws, fire, defoliants and huge machines. Due to the fragile nature of the tropical soils and the intense heat of the sun, little of this will have the chance to regrow in its original state and vast land areas could quickly end up as scrub or desert.

The consequences for people living in the tropics are immediate and catastrophic. For the millions of aboriginal people in the forests of South America and elsewhere, forest loss means loss of homes, food, and security, contact with outsiders who bring diseases to which they have no resistance, and frequently cruel repression – even genocide. For the poor and landless people who are forced into forest regions to try to farm, loss of tree cover exposes the soil to intense heat and monsoon rains, bringing soil erosion and flooding. Tropical soils are often exhausted after being farmed for two or three years, whereupon the farms are usually abandoned, there being no money available for fertilizers.

In the longer term, deforestation in the tropics is bad for us all. Loss of tropical rainforests threatens literally millions of plants and animals with extinction, wastes potentially vast resources of food and medical drugs and reduces the overall genetic stability of the planet. Huge supplies of raw materials are also squandered during haphazard and wasteful forest clearance. And deforestation on the scale now in progress has frightening long-term effects on climate, because trees play a vital role in hydrological systems and in determining carbon dioxide levels, which themselves influence the global temperature.

However, we do not have to look as far as the tropics to see widescale forest damage. In the developed north, our own forest resources are frequently being squandered with little thought of the future; replanting all too often comprises single-species plantations which have few benefits for ecology and are less able to withstand the environmental pressures of climate and disease.

In the temperate regions, it is not only felling that is causing problems. Intensive forestry management and the global trade in timber have helped spread many tree diseases, and whole species are now threatened with the fate of the English elm, through fungal and parasitic attack. Plantation forestry of the type practised by Britain may yet prove to be unworkable in the long term, making the forests more liable to damage from bad weather conditions and disease, and exhausting soil fertility by intensive 'cash cropping' rather than sustainable management. And over the past few years, the effects of pollution on forests have suddenly become front page news, with huge areas now suffering damage in Central Europe and North East America. In Europe a total area about half the size of Austria is now thought to be affected.

Britain lost most of its great forests hundreds of years ago, but during recent decades there has been a sudden acceleration in the decline. Since the Second World War, over half the remaining natural or ancient woodlands have been felled, often to be replaced by exotic plantation species or simply cleared to allow bigger agricultural fields. Hedges, which have frequently provided an invaluable woodland substitute for wildlife, are being ripped up at a frightening rate to allow bigger farm machinery to be used; some 150,000 miles have been lost since the Second World War, and farmers in parts of Britain are now facing serious soil erosion problems as a result of losing their windbreaks.

The forests that do remain in Britain are often slowly degenerating under present management policies. In the Snowdonia National Park over 90 per cent of the woodlands are failing to regenerate because sheep are eating the young saplings and preventing regrowth. The price of energy and the increased use of woodstoves has done further damage, with people frequently taking wood for fuel without replanting trees.

Whilst environmentalists have been protesting about forest destruction for some time, the emphasis has usually been on the rate of forest loss and the actual mechanisms of felling (for farming, firewood, timber and so on) rather than the root causes of the situation. Forest destruction is not a chance occurrence or inevitable development. Nor is it simply the result of population growth or rural development. Until recently, the huge jungle areas were virtually impenetrable, and protected by their very inhospitality. Modern technology, with aeroplanes, power machinery and antibiotics, has opened up the forests and, once accessible, they have quickly become one more resource to be exploited, frequently by companies based in the richer countries. Forests have not only become a cheap answer to the 'problem' of landless people who can be settled there, but they are also a dump for the pollutant by-products of industrialization. In addition, forests themselves are a powerful symbol to many people in both the Third World and in the north, and their clearance often has as much to do with personal and national pride as it has to do with straightforward profit.

Whilst it is sometimes tempting to lay the blame for forest problems completely on the greed of individuals and corporations, it would be a dangerous oversimplification to ignore the very real effects of growing populations and, crucially, the political development of the countries possessing large remaining forest areas, a statement that is as true for northern countries as it is for the south.

This book attempts to show the significance of what is happening to forests, and how the developed countries have acted with blinkered and reactionary interests in developing countries to bring about the critical situation we face today. Forest loss is not just the concern of a few middle class nature lovers. It has urgent implications for everyone, and especially for those who will inherit the legacy of ruined land left behind by the people who are currently reaping the profits of destructive exploitation.

1. The world's forests

Forests currently cover about one fifth of the earth's land surface, although in the past the forested area has been far larger. Understanding forest issues is easier with at least a general knowledge about types and distribution of present-day forests. The two main types are temperate forest and tropical forest.

Temperate forest

Temperate forests are found in land areas which are warm enough and low enough to support trees but not so hot as to be tropical. There is a large grey area of 'sub-tropical' or 'semi-tropical' forests in cooler parts of the Third World; for the purpose of this book temperate forests are defined as those in North America, Europe and the cooler parts of Australasia, while the sub-tropical forests of the Third World are included in the section on tropical forest.

To a large extent, the distribution of trees in temperate regions was determined during the last Ice Age. The great ice sheets that covered much of the planet caused large-scale migrations of tree species as the ice slowly advanced, so that some species were cut off in a few hills and valleys, and many gradually became extinct. As the ice retreated, a few species expanded their range again so that they now cover entire continents, while others remain confined to a single mountain or valley.

The ice had very different effects on the Old and New Worlds. In the Americas the mountain chain runs from north to south and trees could 'retreat' along this at their preferred altitude to survive the Ice Age in the more temperate climate of Mexico. In Europe, on the other hand, the mountains run from east to west, so that the trees could only retreat through a few escape routes; many species must have perished in the ice. Britain has even fewer native species than the mainland of Europe because it was cut off more competely by ice; there are only 35 naturally occurring species of tree (although there are many more 'bush' species like the smaller willows). A tropical forest country might have over ten times this number of species.

The Ice Ages and the gradual separation of the continents has meant that there are considerable differences in the composition of tree species between the Old and New Worlds and the northern and southern hemispheres. Although there are tree species in the same genera in both Eurasia and the Americas there is only one species of tree found naturally in both, the common juniper. There are no species (and very few genera) found naturally in both hemispheres.

There are a number of very different types of forest listed under the heading 'temperate':

Deciduous woodland, which tends to dominate lowland areas and includes extensive underscrub and herbs as well as trees. Species vary as a result of a number of factors, including altitude, wetness and past history.

Coniferous woodland, which is generally found on higher ground and has fewer species. Much of the coniferous forest today, especially in Europe and Australasia, is made up of non-native species or is extensively managed.

Mediterranean woodland, which includes both conifers and broad-leaves, but usually contains distinct enough species to be classified separately. This is generally very open woodland, found in the Mediterranean countries of Europe and North Africa.

Temperate rainforest, which is found in parts of North America and Australia and in northern Asia. It is a very rich habitat when compared to drier temperate forest.

These 'types' are not exclusive, and all kinds of mixtures occur in nature.

Tropical forest

Tropical forests cover about 10 per cent of the world's dry land surface, comprising some 900 million hectares (or 2.2 billion acres). Tropical forests are divided between South America (58 per cent), Africa (19 per cent) and Asia and Oceania (23 per cent). Four main geographical regions are identified:

American tropical rainforest region, which includes parts of Central and South America and the Caribbean.

African tropical forest region, which includes the Congo basin, the coastlands of West Africa, the Cameroon highlands and the forest of Madagascar.

Indomalaysian tropical forest region, which includes parts of India, Burma, the Malay peninsula and the South East Asian islands.

Australian tropical forest region, which includes North East Australia, Papua New Guinea and the surrounding islands.

Hawaiian islands, which form a fifth small group.

The main countries possessing tropical forests are Brazil, which contains almost 33 per cent of the total, Zaire and Indonesia which each have 10 per cent, Papua New Guinea, Colombia, Venezuela, Peru, Gabon and Burma which each contain over 200,000 square kilometres of tropical forest. Thus Brazil, Zaire and Indonesia jointly own more than half the world's tropical forests.

It should not be imagined that tropical forests are similar throughout the world, any more than deciduous forests are. In fact the variation in tropical forest is much greater. 'Tropical forest' is a general term for forests found in tropical monsoon regions of the world; within this area there is a wide variety of forest types including tropical moist forest, cloud forest, swamp forest, coastal mangrove and deciduous woodland, all with their own geographical and climatic characteristics.

Two thirds of tropical forests are tropical moist forest or rainforest, and the remaining third is moist deciduous forest. As its name suggests, rainforest is wetter, with 4,000–10,000 millimetres of rain per year, as opposed to 1,000–4,000 millimetres in moist deciduous forests. (Compare this with 6-800 millimetres in London, for example.) Tropical rainforest is usually richer in both number and variety of species.

The main distinguishing feature of tropical forests is their enormous age. Tropical forests are extremely ancient and, if undisturbed by humans, very stable ecosystems. During their long history they have built up an enormous diversity of species, even in a fairly limited area. This includes the existence of 'biogeographical islands' where distinct plant and animal assemblages are found. These rich oases are thought to be areas that were never covered with ice during the last Ice Age and have therefore retained ancient and diverse forms of life as a result. Particularly striking examples of these exist in the Amazon rainforest.

Diversity in animals and smaller plants is mirrored by the trees themselves. Those in tropical forests tend to reproduce very slowly, so that no single species comes to dominate all the others, as is often the case in temperate woodlands. Large numbers of species can apparently coexist indefinitely, each being at a very low population density. In the Amazon jungle over 2,500 species are known and doubtless more are still to be discovered. Up to 400 can be found in a single hectare, whereas in Britain it would be rare to find twenty species in one hectare in a natural woodland.

Despite the luxurious growth in tropical forests, most of their soils are poor or even almost sterile. In temperate woodlands dead plant and animal matter builds up in the soil as humus, so that most of the available nutrients are in the soil rather than in living plants. In the tropics, on the other hand, nutrients tend to be recycled extremely quickly. They are collected from airborne dust, from rain through the leaves, from the ground layer and the surface of the tree. Many tropical trees grow roots out of the ground to collect nutrients from surface humus and out of their branches to take nutrients from the remains of other plants growing on the tree itself. Up to 60 per cent of the available nutrients can actually be locked up in living material, leaving just 40 per cent in the soil.

This extraordinary efficiency is the secret of the jungle's lush fertility. However, it also means that once the plants are removed, the soil very quickly becomes worthless; there is no 'insurance' against removing the vegetation cover. (Of course, if trees are clear-felled often enough in temperate forests the soil will lose fertility, as is happening with intensively managed plantation areas, but the process takes far longer and is more easily correctable.) Light-loving weeds and scrub can quickly invade cleared areas in tropical forests, preventing regrowth of trees.

In other areas, the consequences of removing plants are even more severe. In regions of very heavy monsoon rains, the thin soil has little protection against erosion, and trees have evolved to provide a very dense canopy so that rain reaches the ground as a fine mist rather than hard droplets. Once tree cover has gone, the soil is first baked hard and cracked by the sun, then washed away by heavy rains. As we shall discuss later, this soil erosion is extremely severe, and disastrous for people living in the region. Tropical forests have with accuracy been described as 'deserts covered with trees'.

2. Felling the forests

Perhaps the best collective term for the felling, burning, bad planting, disease and pollution threatening our trees is 'forest abuse'; this term will be used here when the general effects of damaging forests are referred to together. By far the most serious and irrevocable forest abuse taking place today is the very rapid destruction of forests in the Third World, both in tropical jungle areas and in arid scrubland, creating an unparalleled ecological disaster. Although temperate as well as tropical forest abuse will be discussed in this book, it should be stressed that our problems in the north are insignificant compared to the large-scale disasters facing the south. However, as I hope the following pages will show, this is no cause for complacency; the effects of tropical deforestation have implications for us all, wherever we live.

A report prepared in 1980 at the request of Jimmy Carter puts the current deforestation phenomenon succinctly into perspective:

> Twenty-two years ago, forests covered over one fourth of the world's land surface. Now forests cover one fifth. Twenty-two years from now, in the year 2000, forests are expected to have been reduced to one sixth of the land area. The world's forest is likely to stabilize at about one seventh of the world's land area around the year 2030.*

This means that the world is facing a new, and more serious, acceleration of deforestation. More serious both because of the increased rate of destruction, and because forest is being destroyed in areas where its loss is having immediate and often irreversible effects. Deforestation is arguably the most serious environmental

---

* *The Global 2000 Report to the President*, London: Penguin 1980.

issue of the next few decades. But the term 'deforestation' is a hopelessly broad definition. It is worth mentioning here the two main types of deforestation that affect both northern and southern countries:

1. Permanent deforestation, where forests are removed and replaced by farmland and cities, or by scrubland and desert.

2. Temporary deforestation, where trees are removed, ideally selectively, and forest eventually regrows to approximately the same form as before, or forest plantations are established.

These categories are not completely distinct. Areas that are cleared for farming sometimes revert to forest later, or are replanted. 'Temporary' deforestation can effectively be permanent if logging is done insensitively and too many of the remaining trees are damaged beyond chance of future survival. What the ecologist would regard as adequate replacement forest for wildlife or soil conservation may be very different from what the timber merchant will be satisfied with. Nonetheless, the distinctions are important, especially in the tropics where complete removal can easily lead to scrub formation, while selective removal, if practised well, at least gives the chance of regeneration.

Tropical forestry

Loss of trees in the tropics has come to dominate the discussion about deforestation so completely that for many people deforestation effects equal the effects of tropical forest loss. In some ways this is not unreasonable, because tropical forests are so different from temperate woodlands that the effects of felling them cannot be compared directly with our experience in Europe or North America. Many of the problems that the tropics are facing have come about indirectly because western principles of forestry have been assumed to hold true for tropical rainforest as well.

Effects on forest dwellers

Perhaps the most immediate and irreversible effect of felling tropical forest, especially in South America and Africa, is the disastrous consequences it can have for aboriginal* peoples living in forest regions. Tropical forests are the home of many small-scale human societies. These groupings are frequently extremely ancient, self-sufficient entities, living chiefly by hunting and gathering, although some engage in agriculture as well. They are unique in today's world in the extremely complex relationship they have with a particular area of forest, and their intimate knowledge of the forest's ecology, food value and dangers. It is by no means uncommon for children within a tribe to be capable of identifying several hundred species of plant by the age of five or six, and of knowing their food and medical values, their dangers and other properties that are useful to the tribe.

Conrad Gorinsky, an ethno-botanist working in the Amazon, writes:

> People who have been living in tropical forests with no
> written culture have a knowledge which is distinctly
> different from that of western society with its written
> culture. Knowledge of the forest is transmitted from one
> generation to the other through intimate contact with the
> forest. The forest is as immutable as a library.
> (*Undercurrents*, Vol. 47, 1981)

Undisturbed, the forest peoples can coexist indefinitely with the forest, controlling their population size by practising contraception, and via cultural mechanisms such as taboos. There is a great variation in types of tribe; some are peaceful while others are warlike, some patriarchal, others matriarchal, and so on.

After centuries of existence, forest peoples the world over are threatened by the development and felling of forests. Physically unable to cope with new diseases, and socially threatened by the high-consumption capitalist society to which they are abruptly introduced, tribes are disappearing at an enormous rate through disease and persecution. While the trans-Amazonian highway was being built, unknown tribes were encountered at a rate of about one per year, but many of these groupings have already

---

* 'Aboriginal' is used here as a general term for a native grouping with an ancient lifestyle.

disintegrated under the onslaught of westernization.

It should not be imagined that forest peoples always disappear by chance or by cultural pressure. Many Third World governments regard aboriginal tribes as a nuisance, little superior to children or animals, rather than as independent and highly cultured peoples. The persecution of forest dwellers is amongst the greatest and most understated atrocities of the last few decades, many thousands have been killed during this century to make way for mines, ranches and roads. In the Matto Grosso, smallpox, influenza and tuberculosis were deliberately introduced to reduce tribes. In Brazil, tribes have been bombed from the air with dynamite and one group was virtually wiped out when arsenic was added to their food. In Paraguay, government soldiers have hunted tribespeople and shot them to clear areas wanted by business interests.

Even when genocide is not actually practised, some of the alternatives are little better. Many tribes have been forcibly airlifted to new sites, but the changing nature of the forest means that they are often moved to areas where they do not know the food plants, and are frequently unable to survive. In other places Christian missionaries are on record as using 'tame' Indians to lead tribes back to mission stations where they are kept against their will, and where many die of disease, starvation or shock.

Some of the worst abuses have taken place in South America, where some two million Indians have disappeared in Brazil alone during this century. Persecution also takes place in Africa and Asia. In the Philippines, Roman Catholic bishop Francisco Claver said in a letter to President Marcos that the persecuted Tingian tribe 'seem to have a far richer and more comprehensive notion of human development than that espoused by the government . . . What the government defines as development for them is actually the destruction of all that they value in life, above all their spirit as a people. This spirit is no small thing for survival itself.'

The sufferings of aboriginal peoples have not gone unnoticed of course, even if there has been a generally weak response from other countries. As a result of international pressure, considerable lip service is now paid to Indian rights by most governments in tropical forest countries. Several have set up special groups to deal with the 'problem'. Unfortunately, reality is often very different, and aboriginals are still disappearing. Protection groups are under-financed, poorly staffed, and often extremely corrupt; they are unable to halt the atrocities or the more subtle pressures even if they want to. Paradoxically, the creation of national parks to protect tropical forest areas, a development normally welcomed by progressive interests, has sometimes meant that native peoples have been expelled from their home area in order to 'protect wildlife'.

The blame for native peoples' problems should not be confined solely to the governments of their home countries. The poverty caused to some countries by the present economic order is doing little to improve the aboriginals' chances, because even if governments wish to help, they frequently have very limited funds to do so. Nonetheless, the destruction of forests remains the chief threat to many forest dwellers.

Effects on wildlife

It is well known that tropical forest areas are particularly rich in wildlife. It is worth emphasizing this diversity and comparing it with other regions.

Although the biological composition of jungle areas is still scarcely known, scientists believe that up to five million species may exist in tropical moist forest. At present about 20 new species of insect are discovered every day somewhere in the world, along with 15 species of plant. There are not infrequent encounters, also, with unknown fish, birds and even mammals. Most of these species are found in tropical forest habitats. A recent expedition to the Panama/Colombian border discovered almost 50 new plant species on a single mountain, while one volcano in the Philippines had more woody plant species growing on its slopes than in the entire USA mainland.

Ecologist Norman Myers, who has done as much as anyone to draw attention to the plight of tropical forests, writes:

Costa Rica, a country comprising slightly more than 50,000 square kilometres (hardly bigger than Denmark and a little less than West Virginia) has 758 bird species, 620 of them resident, or more than are found in all of North America north of the Tropic of Cancer. Costa Rica likewise harbours over 8,000 plant species, with more than 1,000 orchids. The

La Selva reserve, amounting to only 730 hectares, contains 320 tree species, 42 fish, 394 birds, 104 mammals (of which 62 are bats), 76 reptiles, 46 amphibians and 143 butterflies – a tally that is, broadly speaking, half as many again as California's. (*Ambio*, Vol. 10, 1981.)

However, this incredible diversity also brings with it fragility once the delicate fabric of the forest is destroyed. Because of the large variety of species, comparatively few of each individual species are found (Amazonian trees often exist at levels of only about four per hectare), so that species are far less resilient to damage than their relatives in temperate forests, a situation that is made worse because many species are only found in a very limited area. Even a relatively small amount of felling can destroy (and is destroying) an entire species or even a whole group of species.

Loss of species by extinction (i.e. their complete disappearance forever) is thought to run at about one per day at the moment, although no one knows for sure. Most of these will be small plants and animals, but larger species are already increasingly at risk. Norman Myers draws this horrifying scenario:

Let us suppose that . . . as a consequence of manhandling of natural events, the final quarter of the century witnesses the elimination of a million species – a far from unlikely prospect. This would work out . . . at an average of 100 species per day . . . By the time human communities established ecologically sound lifestyles, the fallout of species could total several million. This would amount to a biological debacle greater than all the mass extinctions of the past together.*

Larger and more colourful species are further threatened by the wildlife trade, when forest areas are opened up to exploitation. Literally millions of plants and animals pass from Third World countries to Europe, North America, Japan and the Middle East every year, to be sold as pets, or to provide expensive clothing, luxury food – even aphrodisiacs. The demand for these items is an

* Norman Myers, *The Sinking Ark* Oxford: Pergamon Press 1980.

overpowering incentive to many poor Third World dwellers, and even where national forest parks exist they are all too often heavily poached for the most valuable species.

At present, the prospects for wild species in the tropical forests look surprisingly grim, despite the platitudes from the north over the last few years. With extinction so easy, and long-term protection virtually impossible to ensure, the chances are that there will be a steadily increasing trickle of disappearing species over the next few decades.

Endangered species include many animals and birds as well as smaller organisms. Of the 30 or so primates believed to be facing serious risk of extinction, the majority are forest dwellers. Large cats, crocodilians, snakes, numerous other reptiles and fish are all threatened when the forests disappear.

It may seem indulgent to worry about plants and animals, even if they are disappearing on such a large scale, when Third World countries already face such huge human problems. However, wild plants and animals play a vital role in many very pragmatic ways, quite apart from any moral compunction we might feel for them. Below we discuss just two of these, their use as drugs and as food.

Effects on medicine
Forty per cent of drugs commonly used in the north today were originally derived from wild plants. This does not include specifically herbal remedies, nor does it take account of the thousands of plants used by local healers in Third World countries.

Of these plant-based drugs, some of our most spectacularly successful have come from the tropical forests. Quinine made settlement in the tropics possible by combatting malaria. It is synthesized from cinchoma bark. Curare was in many ways the starting point of modern surgery through its anaesthetic properties. Its properties were learned from Indians who used it to tip poison darts for hunting. It is now in short supply, because it is being harvested from vines when they are still too young, and also because the Amerindians who are skilled in finding it are rapidly disappearing.

The current birth control pill was originally synthesized from the Mexican yam, another tropical plant. More recently, the

Madagascan periwinkle, found in an area of rapidly dwindling rainforest, has yielded two drugs important in fighting Hodgkin's disease and cancer, while a survey of 1,500 Costa Rican plants found that as many as 15 per cent had potential anti-cancer properties.

Harvard botanist Richard Evans Shules believes that discoveries like this will trigger a revolution in medical attitudes to forests and genetic resources:

> It crystallized the realization that the plant kingdom
> represents a virtually untapped reservoir of new chemical
> compounds, many extraordinarily biodynamic, some
> providing novel bases on which the synthetic chemist may
> build even more interesting structures.*

We know that many aboriginal tribes in rainforest areas have been using natural plant drugs successfully for hundreds if not thousands of years. The World Health Organization has recently started screening plants for useful properties, but at the same time the very existence of many plants is threatened and the knowledge built up gradually by Indians is being destroyed along with their culture. There is undoubtedly far more that we can learn from tropical plants and people, but time is running out while the forests are disappearing.

Effects on food

In much the same way, we find that many of our staple food crops originated in forest areas. The potato and cassava are both believed to have been discovered by Amerindians in Latin America, along with more specialized food like tomatoes, peanuts, Brazil nuts, cocoa and cashews. Many grains, fruits and vegetables are thought to be originally forest species.

However, despite the enormous diversity of plants, we have come to rely on very few species for our food crops; today 85 per cent of the world's food comes from just eight plant species. Humans have never used more than about 3,000 species in all, yet

---

* Robert and Christine Prescott-Allen, *Genes from the Wild*, London: Earthscan, 1984.

scientists estimate that there are at least 70,000 other possible food plants. Some of the reasons for this are historical and genetic; certain plants like rice and wheat have been bred to give a high yield in a wide range of situations. However, the spread of food species is increasingly controlled by agribusiness companies who have a vested interest in the use of certain crops for which they supply the fertilizers and herbicides, and which they are equipped to process and market.

Pressure from food transnationals has been felt throughout the world during the last few decades. It appears in Europe in the guise of draconian EEC regulations which limit the varieties of fruit and vegetables that can be sold in Europe to those for which there are chemical sprays. And in the Third World there has been a continual move to limit crops to those which have a ready market in the north (decided upon and controlled by the transnationals), eliminating local varieties.

This puts agriculture on a very precarious footing indeed. If any crop plant succumbs to a devastating disease (like the potato blight which helped cause the Irish famine in the ninteenth century, or Dutch Elm Disease which defeated the best efforts of arboriculturists only ten years ago), the effects on humans would be devastating. And by putting all our nutritional eggs in one basket, we are at best producing food extremely inefficiently, in that crop species are having to cope with a wide variety of habitats and climates, bolstered by the expensive application of chemicals.

Tropical forests can play an important role in helping to stabilize the world's food supply, if they are allowed to survive long enough to do so. Forest plants can fulfil three main roles. Firstly the forests can continue to do as they have done in the past, namely serve as a source of new food materials for breeding and cultivation. Where aboriginals' knowledge of food plants has been used, the results have sometimes been spectacularly successful. The winged bean was only known to a few Papua New Guinean tribes until 1970, but is now serving as a high-protein crop in over 50 countries. The wax gourd vegetable *Beniscasa hispida* grows in the Asian tropics, reaches 35 kilograms in weight and grows up to two metres long. It can be eaten as a cooked vegetable, as sweet or soup stock and, uniquely, it can be stored for up to a year because a tough, waxy coating prevents it from being attacked by micro-organisms. Around 125 fruit species are already cultivated from the forests of South East Asia; at least the same number again are available for development. Similar abundance occurs in Latin America, including fruits with names like *feijoa*, *naranjilla*, *jabaticaba* and the tree tomato. At least 1,650 leafy plants in tropical forests contain as much protein as legumes and fruits in addition to vitamins and many important trace elements. Millions of people in developing countries suffer lack of vitamin A because of a scarcity of milk and eggs, but this could often be supplied by plants growing on their own doorstep.

Second, the continued existence of forests is essential to species which we have already cultivated. Contrary to general assumptions, 'domesticated' species are not static organisms but are, like wild species, constantly adapting to fit changing environmental conditions. However, in the case of crop species the 'adaptation' is usually carried out fairly artificially by crop scientists. Commercial strains can have a lifetime of just a few years before succumbing to inbreeding or pest attack; the pests themselves are constantly changing in response to new pesticides and can rapidly adapt to attack previously 'safe' crop strains. When this happens, one common method of improving a crop is by interbreeding the cultivated strain with its wild cousin, to inject 'hybrid vigour' into the crop plant. The continued presence of wild species is needed to maintain the health of the crops we already possess as well as providing a resource of new crops.

Third, wild genes are already widely used in many of the world's most important crop plants. Several wheat cultivars have obtained fungal resistance through breeding with wild strains. This has also provided the wheat with drought resistance, winter hardiness and heat tolerance. Rice acquires resistance to two of Asia's four main rice diseases from a single sample of central Indian wild rice; its use for breeding has helped double Indonesia's rice production. Genes from wild maize in Mexico, wild cassava in Nigeria, wild oil palms from Africa, wild potatoes from Latin America, wild tomatoes, wild cotton and wild sugar cane are all providing essential disease resistance which crop scientists are still unable to produce artificially.

Perhaps the most useful long-term nutritional role for forests is to provide us with a food source by harvesting plants in their present state. The scope for traditional, settled farming on poor tropical soils is often very limited, given restrictions of geography and economics. Several experimental systems have instead used the forests as a food source directly, by cropping a wide variety of species in a sustainable manner in much the same way as forest aboriginals do, perhaps with some selective planting to maximize food production.

It is not just plant resources that are at risk if the forests continue to disappear. Resources consisting of fish and mammals have scarcely been exploited at all, and native cattle species are thought to be far more resistant to indigenous diseases such as trypano-somiasis (sleeping sickness, carried by the tse-tse fly) than western cattle imported into a radically different environment.

Conrad Gorinsky is convinced that aboriginals hold the key to harvesting many future foods:

> Scientists who have made a study of the knowledge of the Amerindians are amazed that they can identify every single tree in their environment . . . it is obvious that the knowledge these people have on the identification and use of plants is enormous. But it is orally held and therefore there is an extreme danger that this knowledge is being lost because there is no way of securing the culture of the forest people.*

Effects on the gene pool

As our discussion about crops has shown, part of the importance of the forest areas, and particularly the immensely rich tropical forests, is that they serve as a source of raw material for crop breeding and adaptation. Another concept that has gained widespread acceptance in the last few years is that the huge numbers of species in the tropics also provide a reservoir of plants and animals available to adapt to future changes in environmental conditions. This 'gene pool' is now seen by many scientists as essential to the continued evolutionary stability of the planet. This was recently listed as the second most important environmental

---

* Interview in *Resurgence* magazine, 1982.

research priority for the 1980s by the Swedish National Academy of Sciences journal *Ambio*; the top priority was to combat depletion of tropical forests. If the rate of extinction develops as fast as seems likely, we may well see a major disruption in many of the evolutionary processes that have developed over millions of years.

The fact that the rainforest is so rich means that it will inevitably become the major gene bank for the future, both to fulfil human needs and as a source of genetic raw material for the planet. Each time a species of plant or animal becomes extinct the world loses genes that can never be replaced; we have reduced our evolutionary options for the future. The rapid rise of biotechnology is opening up a potentially huge market for raw materials produced through evolution, and the forests are among the richest hunting grounds in the world for these.

The World Conservation Strategy, in its report of 1982, sums up the importance of the gene pool as follows:

The preservation of genetic diversity is both a matter of insurance and investment – necessary to sustain and improve agricultural, forestry and fisheries production, to keep open future options, as a buffer against harmful environmental change, and as the raw material for much scientific and industrial innovation – and a matter of moral principle.

The present attitude to preservation of species is often that a few nature reserves can act as indefinite reservoirs or, at least, can preserve species until larger forest areas are restored. Unfortunately, such a simplistic attitude is unlikely to produce satisfactory results. If a genetically rich (i.e. species rich) area of forest is just cordoned off and the trees around it felled, it is unlikely that all the species initially inside it will survive very long, however 'undisturbed' it may remain, unless the area is very sizeable.

Exhaustive work on the ecology of islands (which always have fairly low total species numbers) has shown that small areas have fewer species than large areas, even if there is theoretically space for them all to survive. Unless they are artificially managed, species become extinct in small nature reserves. (Therefore one large nature reserve is better than a few small ones adding up to the same total area.) If the gene pool is to be preserved in the tropics, we need large areas to be preserved, preferably those which already have the largest diversity of species.

Effects on soil erosion
The main occupation of people living in tropical forests is now agriculture. Unsuitable though forests are for farming, they have provided an ideal area to relocate unwanted peasant people who are without land of their own, either because land ownership is concentrated in the hands of a few people, or because expanding populations are forcing people onto less suitable land. Religious and political persecution, drought, disease, famine and concentration of capital all add to the growing numbers of landless and migratory people who end up in the vast moist and dry tropical forest regions of the world.

Once there, they face the colossal problems of disease, climate, soil infertility and, frequently, official indifference to their plight once they have been shunted out of sight. For them, the immediate effect of deforestation is the massive soil erosion that can result both from their own agricultural efforts and from deforestation due to logging or firewood collection nearby. Both land and people suffer and, once started, soil erosion is extremely difficult to halt or reverse; it leads to permanent impoverishment of huge land areas.

As soon as the protective mantle of trees has been removed, the soil is ripe for erosion. It dries and cracks in the hot sun and torrential monsoon rains wash it into rivers and into the sea. Water courses become silted up, low-lying land is flooded and ruined, and river navigation is blocked. Remaining trees are undermined and any crops being grown are spoiled, leading to local food shortages which further increase the problems for those left without homes or farms. In just a few short years the habitat of a deforested area can be altered forever.

Damage can be quite extraordinarily fast. Tropical forests can often receive as much rain in an hour as London would expect in a wet month, and a single storm has been measured as removing 185 tonnes of topsoil per hectare. The land often has no second chance after deforestation.

Even aside from such huge water erosion problems, tropical

soils face other threats from the loss of tree cover. Once soil temperature exceeds 25° centigrade, humus decomposes faster than it forms; volatile nutrient ingredients like nitrogen can be lost, further reducing the fertility of the remaining soil. Winds can also remove topsoil and attempted ploughing can damage the soil structure. While soil loss has immediate effects on those farming in the area, it can also cause problems further afield. Silting up of rivers damages fishing and shipping, and can block reservoirs and irrigation schemes. In Argentina, $2 billion has to be spent every year to dredge the harbour at Buenos Aires, in order to allow vital shipping to reach port. The offending silt comes from the River Plate, and 80 per cent of this silt reaches the river from just 4 per cent of its catchment, where deforestation has taken place. As forest loss increases, some harbours will face closure because of uncontrollable siltation problems.

Many of the massive hydro-electric schemes favoured by Third World governments as 'long-term' energy sources are only lasting a fraction of their supposed lifetime because of siltation through soil erosion. The life expectancy of the Ambuklao dam in the Philippines has been reduced from 60 to 32 years because of deforestation effects. Electricity rationing has already been introduced into Colombia and Costa Rica because of reduced power supply as a result of deforestation near hydro-electric projects.

In many parts of Asia, soil erosion is further speeded up by burning cattle dung when no more timber is available, thus wasting a source of organic matter which could help stabilize the soil, as well as losing the potential fertilization value of the dung. The practice of patting the dung into cakes for drying also spreads diseases like typhoid and cholera, along with many eye infections.

Throughout the world, deserts are on the move again, spreading quickly in the wake of forest clearance and overgrazing. It is now thought that several 'natural' deserts, especially in Asia, were actually once fertile areas and that desertification was brought about by past deforestation – a situation analagous to the creation of dust bowls in Canadian prairies. Once deserts are established, it is difficult to get rid of them, especially in the inhospitable monsoon areas of the tropics, and as trees cycle a great deal of water, removing them means that the climate changes, so that water is no longer available to bring vegetation back to the area.

The Brazilian government now admits that much of the land cleared for 'agriculture' in the Amazon has proved completely unsuitable. Unfortunately, land clearance continues throughout forest regions, as people with nowhere else to go continue to try and farm their 'desert covered with trees'.

Effects on climate

Of all the possible effects of deforestation, perhaps the most potentially serious are the changes in climate which are brought about by loss of trees. However, this is the area with the most radical disagreements between experts, so no consensus of opinion is possible. Possibilities have ranged from scare stories that all our oxygen will run out, (it won't), to the very real threat of disruptions in local and global climate. There is no overall agreement about global effects as yet, but many climatologists are now seriously alarmed by the implications of continued forest loss.

Local climatic effects are undisputed. Trees play an integral part in local climate by passing groundwater to the atmosphere through their 'transpiration screen' via roots, trunk and leaves. They also stabilize the soil and help diffuse the falling rain with their leaf canopy. Trees reduce soil temperature and increase local humidity. Whilst this is known in general terms, the precise consequences of felling trees are still unclear, and will vary from place to place. The destruction of tropical forest tends to decrease rainfall in the areas cleared. This reduction is at least partly due to the lack of transpiration; half the rainfall in the Amazon is believed to form in this way. The removal of green trees also exposes a paler soil (especially once it has dried out) and an 'albedo effect' reflects more heat and light back into the atmosphere than would be the case if the sun shone on green trees, causing further climatic changes.

Possible global effects are less completely understood, but may perhaps turn out to be far more significant. Cutting and burning forests, which are reservoirs of stored carbon, adds to the amount of carbon dioxide in the atmosphere. This is already increasing in concentration because of the burning of fossil fuels. By 2020, if trends continue, the level of carbon dioxide will have doubled from that which was found before the Industrial Revolution.

Although carbon dioxide makes up only a very small part of the atmosphere (about 0.03 per cent) it plays a vital role in determining surface temperature through the so-called 'greenhouse effect'. Carbon dioxide molecules in the atmosphere trap some of the reflected heat (which is at a different wavelength to incoming solar radiation) so that an increase in atmospheric carbon dioxide leads to a rise in global temperature.

Some climatologists claim that a very small increase, say of 1.5-3° centigrade on average, could result in the melting of the Western Arctic Icecap and in the increase of sea levels, on average, of up to 5 metres, inundating low-lying areas like Florida and the Netherlands, and swamping many ports. Historical evidence suggests that average global temperatures over the last Ice Age changed by only about 5° centigrade, which lends an uncomfortable credibility to the theory.

Another, perhaps more likely, effect of a temperature rise would be a shift in average agricultural productivity. A global rise in temperature would not be evenly distributed, and comparison with historical data suggests that North America and the grainlands of the USSR would be drier, while North and East Africa, the Middle East, India, Mexico and Western Australia would all be wetter. Thus the USA would produce less grain while other areas, like Siberia, would produce more. At present America's grain surplus is an insurance against global famine, and politically very important. A change in productivity might be accompanied by a change in political power, probably towards Third World countries. However, before this happened, the climatic fluctuations leading up to any shift would almost certainly throw food production into complete disarray, and many people would suffer food shortages.

Some people believe artificial climatic fluctuations are already playing an important role in world affairs, pointing to the apparent increase in the frequency of droughts and extremes of temperature that many areas have been experiencing. It is enormously difficult to distinguish natural fluctuations from human interference, but historical records suggest that areas that have been deforested in the past (like large areas of the Middle East and Africa) must have had a radically different climate within historical times; some of the changes are very likely to have been produced by the removal of vegetation.

The problem with detecting artificial climatic changes is that they take place against a backdrop of natural fluctuations, due to periodic temperature changes, varying output from the sun and many other factors. However, it is a false and dangerous assumption to use the complexity of the issue as an excuse for ignoring it. The effects of atmospheric carbon dioxide simply cannot be predicted at present. But, as one eminent climatologist has pointed out, by the time we do have conclusive proof, it will be 20 years too late to do anything about it.

Destruction of trees in the tropics

Given that deforestation is a generally bad state of affairs, what are the causes of forest loss? Generally, deforestation mechanisms can be divided into two main types, those which revolve around the lifestyles of people living in the area, principally collection of firewood for energy and clearance of trees for farming, and those which involve business interests from other parts of the country or the world, namely logging, mining and ranching. Of the two, fuelwood and farming are affecting much larger areas, but are themselves often dependent on the opening up of forests by commercial interests and the political pressures which deprive people of better sources of land and energy. In practice, the links between the various 'causes' of deforestation are often very close indeed.

Deforestation by fuelwood collection

There is already a very serious energy crisis in much of the Third World, making our own worries about the future of coal and oil look fairly insignificant in comparison. For about 2,000 million people in the poorest nations, firewood is still the main or sole source of energy and wood fuel accounts for about a seventh of total world energy supply. As population grows and more land becomes unavailable, forests can no longer supply all that is needed.

In most tropical forest areas a hectare of woodland can probably keep one person supplied with fuel fairly indefinitely. Today, as many as 15 people will be collecting from a single hectare, with the inevitable result that woodland is being destroyed at an accelerating and disastrous rate. It is estimated that 150 million cubic metres of wood are used for energy production every year worldwide, grossly impoverishing 25,000 square kilometres of primary forest in the process.

According to Erik Eckholm, who has travelled extensively in South East Asia carrying out research for the Worldwatch Institute,

> Trees are becoming scarce in the most unlikely places. In some of the most remote villages in the world, deep in the once heavily wooded Himalayan foothills of Nepal, journeying out to gather firewood and fodder is now an entire day's task. Just one generation ago the same expedition required no more than an hour or two.

To make matters worse, most fuel wood is burnt on inefficient open fires, or converted into charcoal for use in the cities, so that far more is needed than would be the case if efficient woodstoves were used. These are completely unavailable to most rural people simply because of the cost. Open fires are often only 10 or 20 per cent efficient, but without practical alternatives people are burning whatever timber they can get. Even trees which are not actually felled frequently have branches ripped away, and saplings in new plantations are likely to be poached within days of planting.

Social and political attitudes make it difficult to get out of the deforestation spiral. Lack of money means that good woodstoves are unavailable over much of Asia. Community methane digesters, which use dung from livestock and people, have been widely introduced in India, but have run into cultural and caste problems with the reluctance of most people to deal with faeces. Methane digesters also tend to give an intense heat for a relatively short time, ideal for stir-frying Chinese food, which may explain their success in China, but not as useful for the slow simmering needed for Indian food like curry. Many Asian people prefer to use cattle dung, as this gives a slow, steady heat, even though its use spreads disease and impoverishes the land.

Political cynicism and poor communication by government and aid workers frequently ruins attempts to reforest areas. In India, in one instance, villagers planted trees upside down in protest at what they were convinced was just another way of exploiting them as cheap labour, without any real hope of ever using the trees. There is often little awareness about the seriousness of deforestation, despite its scale, and government woodlots are frequently ignored so that the saplings die. Although some countries, including China and South Korea, have embarked on impressive replanting programmes, many Asian countries are still losing far more trees than are being replanted.

Overcollection of firewood is most serious in the dry tropical forests of Asia, where in many cases one member of a family now has a full time job collecting fuel, and cattle dung is substituted when no wood is available. Villages that used to be surrounded by dense jungle within living memory are now alone in a desert or scrubland area with ruined soil and no prospects of future improvement. In the Sahel countries of Africa, deforestation is a historic fact, and it continues today with local timber-burning and charcoal-making for city dwellers. Fuel collection is an important factor in the alarming spread of deserts over the region.

There is evidence that firewood is already so scarce in some areas that it has become a fuel for the rich alone, while the really poor people burn twigs or dung. A study of one Indian village showed that 96 per cent of 'firewood' was actually twigs, while the remaining 4 per cent was branches. On the other hand the city of Bangalore uses nearly half a million tonnes of firewood every year, mostly brought in by diesel trucks which probably use almost as much energy as is contained in the wood.

One of the problems of estimating the effects of firewood collection on deforestation is that official figures often underestimate firewood use. According to one UN official, 'Energy planners of most developing countries are "electricity people", and they are making a deliberate effort to play down the contribution of firewood so that power stations rather than fuelwood plantations will get more funds.'*

Logging and deforestation
Quite separate from the activities of local people with little personal choice in their timber requirements, pressure on forests also comes from commercial loggers interested in timber for

---

*Earthscan Briefing Document, 1984.

export to the developed world. These logging companies are frequently transnational corporations, moving into tropical areas wherever they can, and ripping out timber as fast as possible before political changes make it difficult for them to continue. In addition to foreign investors, many Third World governments themselves exploit forests to provide a cash crop, and timber now ranks as one of the top five exports from the Third World. The amount of tropical hardwood reaching the north has increased by a factor of 16 in the last 30 years, from 4.2 million cubic metres in 1950 to 66 million cubic metres in 1980, with an estimated growth to 95 million cubic metres by 1995.

However, this harvest is neither efficient nor sustainable. Many trees are left needlessly damaged in the scramble for profits, and vast areas of land are devastated by the chainsaw every year. In the Amazon, only about 50 out of the 2,500 species are commercially exploited, even though at least 400 are known to be useful. Felling damages many more trees than are actually taken, especially when thick canopies bind trees together and pull unwanted trees down along with selected timber, so that fungal parasites are likely to attack any damaged trees. Aerial surveys are difficult to carry out because some of the trees are left standing, making it impossible to estimate the extent of the damage at all accurately.

Researchers believe that between 53,000 and 87,000 square kilometres of timber are felled throughout the tropical regions every year by foresters, but vast illegal felling operations mean that even this figure must be an understatement. The Food and Agriculture Organization (FAO) of the United Nations calculates that at least a million square kilometres were leased for timber in tropical moist forests alone between 1958 and 1978.

The use of tropical hardwoods is increasing as timber importers try to find new sales gimmicks for the material. Tropical forests now supply 10 per cent of the world's industrial wood, and FAO predict this will double over the next 20 years. Hardwoods are now used for furniture, double glazing fittings, veneer and even sometimes pulp, in a rush to create new markets. By far the biggest importer of tropical hardwoods is Japan, accounting for 53 per cent of total tropical hardwood imports alone, and bearing responsibility for much of the destruction of the Malaysian peninsula forest as a result.

Many transnational companies are now moving into tropical forest areas, where few take the trouble to replant or extract timber carefully. Governments in many Third World countries are so starved of capital and foreign investment that they have little choice but to let logging continue, despite their awareness of the ecological and social consequences that will arise when the profits have all disappeared. Even when controls are imposed, they are often practically impossible to enforce. Thailand has gone as far as putting the death penalty on timber poaching, but the practice continues. On average 30 forest guards are killed in gun battles with poachers every year.

The facts seem uniformly depressing. Most of Indonesia's low-lying land has now been let to logging corporations, although a study of nine of these companies in 1977 found that none were leaving the required 25 trees per hectare. Half of Malaya's lowland rainforests have already been logged, and the rest are expected to go within the next 10 or 15 years. Most companies prefer to take as much as they can straightaway, rather than risk a change of political direction and the imposition of better controls.

Deforestation and forest farming

Logging often has more serious side effects than simply extracting timber. Loggers build roads, logging companies import workers and their families, thus increasing the local population, and hot on the heels of the timber extractors come landless farmers and labourers, desperate for a place to settle and grow food. So great have the problems of landlessness and international migration become in the past few years that thousands of people are prepared to try farming even in the inhospitable regions of the tropical moist forests, simply because they have nowhere else to go.

The rewards of farming are poor at best and the farmers often end up with worthless land once the fertility has been used up and the soil washed away. Despite all the difficulties, it is estimated that one fifth of the world's tropical forests are inhabited by farmers, and that up to a hectare of forest can be lost to farming for every five cubic metres of wood extracted by loggers. Slash-and-burn agriculture, where a small area of forest is burnt, farmed for a few years, then left to re-establish itself, is a well-established practice amongst many forest peoples, and is a sustainable way of producing food. However, if there are too many people, or if the settlers try to establish permanent farms, the land will quickly start to suffer. Recently arrived settlers, with little knowledge of tropical soils or conditions, tend to act as pioneer fronts, pushing ever further into the forest as land becomes exhausted, and leaving a useless mosaic of empty farms and scrub behind them. Forest farmers currently occupy an area only slightly smaller than Western Europe; much of that land will be permanently ruined for the future.

The problems of attempting forest farming are illustrated all too well by the failure of colonization along Brazil's Trans-Amazonia highways. The Belem-Brasilia highway was completed as long ago as 1959 and was vigorously promoted as an ideal place to settle. Between 1968 and 1978 the population of the area increased from 100,000 to 2.5 million, while the state of Rondonia had an annual population increase of over 20 per cent per year during the 1970s as a result of government incentives following the opening of the Cuiba-Porto Velho highway. However, within a few years the scheme was acknowledged to be a failure and thousands of relocated families were abandoned to eke out a living as well as they could without the state support which they had been promised.

Seen in retrospect, the scheme is a monument to bad planning. Some parts of the road were under water for several months each year during the monsoons, effectively cutting people off from the outside world. No thorough soil studies were undertaken and many areas proved unsuitable for agriculture; the settlers were relying on advisers who were as ignorant of conditions as they were themselves. Tropical diseases have been spread along the highways, and forest clearance is giving other diseases, like malaria, more chance to develop, making the settlers even worse off than before. Now efforts are being made by the government to stop settlement in the Rondonia, but people desperate for land are still drifting into the state in large numbers.

The Brazilians should have learned from experience. A railway laid at the turn of the century to encourage settlement in the forested Bragantina zone certainly brought in a lot of people, but the 30,000 square kilometres of cleared land could not support intensive cultivation and is now a semi-desert.

Forest farming is probably the most important reason for tropical forest loss at present. It is also one of the most difficult to stop, or to pinpoint as the result of any one cause. While the rapidly growing populations in many Third World countries undoubtedly have an effect, other social and political factors are frequently more important in pushing people into the jungle. Failure to recognize and stop these factors will inevitably result in a continued increase in forest farming in the future.

Deforestation and mining

Mining is a relatively minor cause of deforestation but, like logging and ranching, it can be important in opening up areas by road building. It can make them liable to legal or illegal forest farming and consequently to degradation in the future. Mining can also have a disproportionately large effect in some areas, both through the production of huge volumes of effluent and because sites of important minerals are also frequently places where unusual and specially adapted plants are found, so that destroying a small area to dig a mine can threaten rare species with extinction.

Controversy has flared up in the last few years about the plans of the British-based transnational Rio Tinto Zinc to exploit one of the world's largest copper reserves, the Cerro Colorado or Red Mountain in Panama's western Chiriqui province. The mine will cost almost $3 billion to develop, and could yield over 1 billion tonnes of ore. Virtually the whole mountain will be removed. Strong objections have come from the 70,000 Guayami Indians living in the area who have been evicted from many of their traditional lands by coffee and banana growers and are disease-ridden and desperately poor.

The Indians are already struggling to survive by subsistence farming and fishing on marginal land, as well as labouring for plantation owners. They fear that the 27 million tonnes of earth removed every year will damage the quality of the rivers on which they depend and that disease, crime and prostitution will enter the community along with the immigrant mineworkers. However, the government is desperately poor and, although Rio Tinto Zinc is a minority shareholder (49 per cent), its annual turnover is over twice that of the annual government budget and there is little doubt as to who will ultimately control the mine.

Inititial resistance has been led by Roman Catholic priests, but a new leadership is springing up among the young Guayami, who have taken a far more militant line in negotiations, and are determined that their tribe will not suffer the same disastrous decline that has affected others faced with similar developments.

Forest loss in arid regions

If deforestation in the monsoon tropics is likely to be the most serious aspect of tree destruction in the long or medium term, the immediate problems facing people who are forced to live in the arid areas of the world are undoubtedly more acute. In extremely dry conditions, deforestation is a major factor in the current very rapid spread of deserts, which causes loss of cropland and resultant food shortages. Although many social and political factors combine to make deserts, the loss of trees is usually the first physical act of destruction. According to Mostapha Tolba, head of the United Nations Environment Programme, 'There is no doubt that the process of desertification is accelerating, with millions of hectares being lost every year.'*

A letter from a relief worker in Eritrea, in August 1984, describes the situation thus:

Some anecdotes. One of the starkest – though I didn't realize it at the time – came from a village farmer in Rora. He was embarrassed and said, 'These people will laugh at me, I'm ploughing with a cow.' His oxen were all dead because of the drought except for one. In Zagreb an old man was up an olive tree cutting leaves for his cattle – no grass. Just dust. On the ground two oxen could barely manage to scrape a two inch furrow, and they needed frequent resting even then.

Even if the vast areas of existing desert are discounted, something like a third of the Earth's surface is classified as arid or semi-arid. These partial deserts usually receive a maximum of 20 inches of rain a year, although large areas have far less, and rainfall

---

* Alan Grainger, *Desertification*, London: Earthscan 1982.

can drop to zero during the periodic droughts. Millions of people make their homes in these inhospitable regions, living precariously by farming and keeping livestock. Today over half the world's arid regions, comprising some 30 million square kilometres, are under direct threat of 'desertification', which is effectively irreversible degradation into complete desert.

Desertification is far more widespread than is usually imagined by people who live in wet temperate countries. It currently affects some 100 nations, about two thirds of the world's countries, in North and South America, Australasia, Asia, Africa and the Mediterranean. Researchers at the United Nations Environment Programme (UNEP) estimate that the desert at the fringes of the Sahara is spreading at a rate of 1.5 million hectares a year. Imagine areas the size of Czechoslovakia becoming wasteland every ten years! Worldwide, some 200,000 square kilometres, an area the size of Senegal, degrades so much every year that food can no longer be produced economically. At the present rate, UNEP estimates that the world will lose at least a third of its arable lands to deserts by the end of the century.

The formation of deserts is no more a new phenomenon than the loss of forests and, in fact, the two often occur concurrently. It is no coincidence that many of the ancient palaces and temples are found in sandy wastelands; the builders did not choose to settle in the desert – they ruined previously fertile land through mismanagement. Large areas of the Middle East, the Mediterranean and the Sahel in Africa were turned to desert centuries ago, and as recently as the 1930s large-scale soil erosion ('dustbowling') occurred in North America when mechanized agricultural methods were used to exploit otherwise good farmland beyond its capacity to survive. What is happening today is more intense desertification, and a rapid acceleration of effects.

There are three primary factors involved in desert formation:

- deforestation
- bad agricultural practices including overgrazing and over-cultivation
- poor irrigation

Behind these immediate factors lie a whole host of other, more complex reasons, including changes in population, the development of capitalism and the vagaries of climate. Recurring periods of drought are an inevitable factor in many desert areas, especially in parts of Africa and Asia, although they seem to be occurring more frequently than in the immediate past. These cause crops to fail and people to starve and, even worse in the long term, mean that existing livestock will overgraze vegetation to the extent that the fragile habitat can be permanently damaged. Each time a climatic crisis occurs, more land is permanently lost and growing numbers of people are forced on to a smaller space.

Deforestation

Richard St Barbe Baker, founder of the 'Men of the Trees' organization, used to say 'when trees go, deserts come'. The first stage in the degradation of arid land to desert is usually the felling of existing trees. Clearance of woodlands in arid land is currently continuing at a rate of 4 million hectares a year, an area about the size of Japan. By far the worst hit area is Africa, which loses about 2.7 million hectares a year.

Tree felling in many Third World arid lands takes place principally to provide fuel and to clear the land for agriculture. Despite the intense heat which occurs during the day, nights are usually very cold in these regions, and fires are needed for warmth and for cooking. People have traditionally used wood; many people turned back to wood again after the 1973 oil price rises when kerosene, the usual alternative, became too expensive. Today fuelwood shortages are endemic in the arid regions, and firewood collection is a full time job for many women, who may have to make journeys of several hours, two or three times a week, to collect fuel for cooking. Elsewhere, the sale of illegally collected fuelwood has become an important blackmarket trade. Armed gangs are often involved in shoot-outs with forest rangers when they illegally fell timber to sell in the towns and cities.

Clearing the land for agriculture is also a very short-term objective. Without the stability provided by the vast root systems of arid land trees, the soil quickly erodes and the water table falls, leaving the area dry and dusty, and well on the way to becoming desert. Once cleared, fresh pressures are brought to bear on the land. Rising populations encourage more intensive farming of the arable areas; less suitable land is brought into production, while the traditional long fallow periods have been largely abandoned. The intensification of the cash economy has meant partial or complete replacement of drought-resistant crops, like sorghum and millet, with more profitable but less hardy species like groundnuts and cotton. Traditional co-operation between keepers of livestock and arable farmers has broken down as barter systems are replaced with money. Aid projects that dig new wells and provide veterinary services encourage farmers to keep more livestock; but when droughts come the animals die of starvation, rather than thirst, and wreck the land by eating all the available vegetation in the process. Nomadic herders have started to settle, imposing further strains on land which is unsuitable for permanent agriculture.

Poor irrigation schemes (frequently funded by the World Bank and benefiting large construction firms) have also done more harm than good. Far from helping arid areas, they can frequently lead to a build-up of salt if they do not drain properly. Thus plants are poisoned and new deserts created, instead of existing desert being transformed into fertile land. Currently some 500,000 hectares of irrigated land become desert every year, roughly the same amount as is newly irrigated. Deforestation can intensify this, both by forming large salt flats around estuaries and by increasing flooding – hence also increasing the concentration of salts dissolved out of the soil.

Even in the richer countries there is considerable evidence of desert formation. When the Navajo Indians of New Mexico were subjugated and confined to a reservation, the American government encouraged them to become sheep farmers. They proved very successful shepherds but, without training in land management and stocking levels, they built up stocks to impossibly high numbers. In one zone where range specialists calculated that a maximum of 16,000 sheep could be supported, 11,000 Navajos were attempting to keep 140,000 sheep – about ten times the safe limit. Places that a century ago were lush meadows are now barren scrub. While the restoration of dustbowl areas shows that good management can pay dividends, it is estimated that an area the size of Utah is still being badly damaged by overgrazing and is in a 'poor' or 'bad' condition.

However, the real problem of desertification does not lie in the north, where funding and expertise are available to reverse the effects, and where climatic conditions are far less severe. Of all the regions, it is the Sahel which has the worst problems, with chronic political and religious instability to add to the environmental problems. As Alan Grainger of the International Tree Crops Institute has said, 'Each year more people produce less food. The fragile, dry lands deteriorate. *And no-one knows what to do about it.*'

Temperate forests

After the catalogue of disasters that accompanies tropical deforestation, the comparatively modest felling carried out in the temperate zones may seem insignificant. And indeed it is relatively so, both in terms of the scale of deforestation and the effects of tree removal. However, this comparatively optimistic generalization hides a number of important factors which are still acting to damage the long-term prospects for temperate forests.

In Europe, there is now very little 'natural' forest left. The majority has either already been clear-felled and replaced by agriculture or urban development, or has been more or less intensively managed for centuries. The area of forest in parts of Europe is now greater than at times in the past, early ecological and resource problems having stimulated massive replanting programmes. The result is that most of the forest growing in Europe today is at least partly artificial, ranging from naturally regenerating trees over much of Scandinavia to highly standardized monocultural plantations in most of Britain's managed woodlands.

It may well be that the felling of the past has far greater effects today than is generally realized. Large-scale soil erosion was recorded over much of Central Europe a few centuries ago, and scientists in Britain now believe that similar soil losses may have occurred in the UK. British heaths and moors once supported extensive forests, yet the soil is now usually very thin and poor; it is possible that today's soil is just the remains of a much more fertile soil which was washed away when trees were felled. Large-scale soil erosion could explain the rapid silting up that took place in some medieval ports like Chester and Bristol. Deforestation was certainly one of the reasons that many mammals became extinct in Britain. Beavers, wolves, reindeer, moose, brown bears and wild oxen were all found at one time, and while hunting was undoubtedly a major factor in their disappearance, the destruction of their habitat left them far more vulnerable to overhunting and must have accelerated their decline. While we do not know of any bird extinctions that have taken place solely as a result of forest loss, the numbers and proportions of species have been radically altered by felling.

The long history of interference means that the small pockets of ancient or semi-natural woodland, which are all that remains of the original forest in some European countries, are of extreme scientific importance. Unfortunately, these fragments are still continuing to disappear, either being cleared for roads and farmland, or being replaced by faster-growing species in plantations.

There are parallels to the European forest situation in North America and although here the natural woodland is still far more extensive, the current rate of its destruction is faster. There is very little untouched forest in the USA, apart from national parks, although regrowth is extensive. Canada still has large natural forests, although these are also being felled at an accelerating rate.

It is important to stress that, while deforestation is dangerous and harmful in the temperate regions, the effects are far less acute than in the tropics. Failure to recognize this, and attempts to use temperate forestry methods in tropical moist forests, are major causes of the current forest problems facing the Third World.

3. Problems with planting

The traditional British one-inch Ordnance Survey maps used to include distinct symbols for different types of forest: broadleaved, coniferous, and mixed woodland. When the new range of maps was introduced in the 1970s this information was left out and woods were represented by an anonymous green. The decision raised a flurry of objections at the time and some conservationists claimed that it was a blatant attempt to conceal the fact that natural broadleaved woodland was rapidly being replaced by introduced conifers.

Forestry officials in Britain often like to point out that there is a greater area of forest in the UK today than there was a century ago. Indeed, throughout Europe as a whole, more forests have been planted than felled in the last few decades, leading some analysts to conclude that there is no longer a forestry 'problem' on the continent. In some countries there is justification for this; replanting has been carried out sensitively and tree resources are apparently being managed in a sustainable fashion to everyone's benefit. But in others, including Britain, the brave new forestry of the twentieth century bears little resemblance to natural woodland and brings a whole fresh range of problems in its wake.

Single species forestry plantations are little more than cash crops, viewed over a slightly longer time scale than agricultural crops. In Britain the vast majority of modern forest plantations consist of one or two imported tree species, planted in regimented patterns, too close together for any other plants to survive (and rigorously kept free of 'weeds' through chemical application), causing long-term problems of soil health and giving little scope for wildlife or amenity. Far from being welcomed as an essential renewal of forest area, plantation forestry has come to be viewed as a serious environmental problem by many people, and causes considerable ill-feeling in many of the rural areas where it is practised.

Afforestation in temperate areas

Britain provides a good example of 'intensive forestry' of a fairly extreme type. Modern forestry practice developed in an attempt to streamline the production of timber. Traditional methods involved growing trees which would only reach a useful size long after the planter's death; this has little appeal in today's economic climate, especially where land area is limited. Plantations are standardized as far as possible to cut out unnecessary labour. The system of planting and management has been designed for maximum efficiency. A typical oak tree will take 150 years to reach maturity, while a conifer will be full-sized by the time it is 50, thus giving conifers a clear advantage, despite the fact that their wood is of poorer quality.

Typically, in upland or marginal land areas, the afforestation process starts with the construction of new roads to allow access for heavy plant and, eventually, extraction of timber. Next, the ground is ploughed, cutting into the surface layer and turning it over, creating a series of mounds and ditches running in straight lines. Tree seedlings are planted on the mounds, giving them an extra depth of soil and allowing easy access for management. In many areas the ground will be regularly treated with selective herbicides to allow the young trees maximum room for growth, spray being applied either by back-pack sprayers or from the air.

Initially, the trees are planted too close together to survive into adulthood. This is done deliberately, allowing weaker individuals to be thinned out and usually sold as Christmas trees, thus bringing in some revenue within a few years. As the forest grows, more trees are thinned, leaving the remainder to reach their required size as a typical conifer plantation: a dark, impenetrable mass of trees growing in straight lines with nothing on the ground except a thick layer of pine needles.

This practice has developed virtually unchallenged throughout the twentieth century, and foresters sometimes boast that Britain has the most efficient forestry in the world. Like most large-scale operations, this 'efficiency' depends very much on the factors that are taken into account when assessing its effect.

Some objections to intensive forestry
In the drive for maximum efficiency, one of the major changes has been in the types of trees planted. Today in Britain, the vast majority of planted trees are not native at all, but are imported species that have been chosen to provide maximum growth rates, often originating from climates similar to our own, like north-west America. While Britain is unusual in having a very small number of native trees, these introductions have also been carried out in most other countries. Furthermore, these trees are not planted in a mixture (as would occur naturally), but in vast areas of single species, or 'monocultural', plantations.

To some extent, the use of fast-growing species has helped re-establish forestry in areas where it had almost died out, or where deforestation had reached dangerous levels. However, many ecologists today believe that the extent of specialization currently practised is no longer 'efficient', and actually poses considerable risks to forests in the future. The use of exotic species has been undertaken on a very large scale without any proof that they will acclimatize to British conditions. The sitka spruce, which today dominates British forestry, comes from Alaska and was planted in quantity after the First World War. Unfortunately, it has failed to adapt to British conditions as well as was hoped, and foresters now fear that much of it will not prove commercially profitable.

In a natural forest, many pests and diseases are kept in check simply because they have difficulty finding their host tree. In the large plantations of today, any pest insect or fungus has no such problem; once it is established in a single species plantation, all the trees present will be open to attack. Recently the pine beauty moth, formerly confined to native Scots pine, has started attacking lodgepole pine, imported from America, which has resulted in various controversial spraying programmes to keep the problem under control. Furthermore, pests of imported tree species are likely to enter the country with the tree, but will have none of their natural enemies to keep them in check once they are established in a new area.

Imported species do not only cause problems for foresters trying to grow them, they can also damage the soil and water relationships in the surrounding area as well. Many fast-growing species cycle large amounts of water through their roots and leaves (the so-called 'transpiration screen'), and this extra water loss has had serious effects in reservoir catchments in many upland areas which currently supply Britain's industrial and domestic water. In experiments on the Plynlimon hills in mid-Wales, run-off from an unforested catchment was averaging 83 per cent of precipitation, while a fully forested catchment was reduced to 62 per cent. The droughts of the 1970s and 1980s have had far more effect than previous dry periods, partly because the water table in many areas has been reduced by afforestation. The natural forests found in the past were far less densely packed, so water retention was not as great.

The quality of the water is also affected. Large stands of conifer have now been conclusively linked with the acidification of water in heavily forested areas, killing fish in lakes and streams and causing wide-ranging effects to other aquatic life. (Another important factor in acidification is air pollution, or 'acid rain', as discussed in a later section.) In many upland areas, trees have been planted so densely that their branches completely cover small streams; fish are either killed by the acidity, or lose their ability to reproduce. Fish in these areas are not adapted to high acidity levels, as they are in parts of the USA where dense conifer forest occurs naturally.

In the long term, the changes to soil and water also result in a loss of fertility. Repeated use of land (which is often already marginal and therefore low in nutrients) for growing successive tree crops, will eventually necessitate fertilization, adding enormously to the costs of producing timber. The rate of fertility loss is the subject of considerable debate; if forestry is managed well it can be negligible, and some forestry experts believe that fertilization will probably never be needed because wind-blown nutrients will replace those lost. However, the new practice of 'whole tree harvesting', much beloved by people wanting to maximize forest profits, increases this loss considerably. The bulk of nutrients in a tree are likely to be in the leaves and small twigs; if these are taken away as well to provide fuel or timber material, the net output of nutrients can be greater than the land is able to sustain. Several countries have already abandoned whole tree harvesting projects in recognition of this problem.

Plantation forestry is also detrimental to wildlife. Anyone who has attempted to force their way through the dense mass of a conifer stand will quickly realize that the ground is almost totally dead beneath the trees; the lack of light and the toxic effect of several inches of pine needles kills any plants that have escaped the attention of chemical sprays. Although a few birds have undoubtedly benefited from the expansion of coniferous forest, other less common species, like the merlin and the golden plover, have declined sharply due to loss of moorland habitat. While bird species increase in the first few years of a plantation, they later fall away again, and commercial forestry is never allowed to grow old enough to build up the diverse habitat found in natural woodland. Thus, when conifer forest replaces broadleaved woodland, the situation is far worse; a particularly rich habitat is replaced by a particularly poor one.

Forests have also been planted at altitudes which are too high, where the cold and rain reduce yields. It is now doubtful whether many of the forests in Galloway in Scotland will be economic. Wind-blow, the blowing down of trees in gales, has also proved a far greater problem than was expected, partly because trees are planted too close together, so that wind cannot pass between them very easily, and also because of the action of ploughing up the land before planting, which can weaken the soil's ability to hold the roots.

For many people the real effect of this kind of regimented forestry is on the amenity aspects of British woodland. Square blocks of trees planted in straight rows look ugly, spoiling the aesthetics of many areas that were previously among the beauty spots of Britain. They bar the way for walkers and other country users. Although the Forestry Commission has made considerable efforts to provide visitor centres and footpaths, the many private forest areas are often virtually impenetrable for countryside users.

Why plantation forestry?

With all the objections, why has plantation forestry grown to such an extent? Initially the Forestry Commission was set up to provide a strategic reserve of timber against the possibility of loss of supplies from abroad during time of war. Although this is still partly the objective of the current Commission, it has also been argued for some years that Britain should be planting stocks of timber now to plan ahead for a future when timber stocks will be in far shorter supply and prices will rise steeply. On the face of it this is a laudable aim, because forests certainly do need to be renewed, but in practice the real reasons for forest development have become buried in a welter of financial and political side-issues, which have in turn adversely affected the type of forest grown.

Financially, forestry is a very doubtful proposition. Although nationalized industries in Britain are expected to show at least a 5 per cent return, the Forestry Commission had this changed to only 3 per cent a few years ago, in recognition of its inherent difficulties in meeting market requirements. Private forestry is usually slightly more efficient, mainly because independent forestry organizations are frequently less concerned with the environmental impact of their work, and have fewer of the statutory obligations to provide visitor centres, nature trails and, increasingly, landscaping of forest areas. While subsidizing forests would be a good and legitimate use of public funds, the poor economics are significant when forestry developments are examined in the light of current tax incentives for what are often badly planned and managed plantations.

The current Conservative government has extended its drive for privatization to the forestry business as well, and there have even been rumours that the Forestry Commission will be disbanded altogether. Although this seems unlikely at the present time, the Forestry Commission is certainly buying less land, and has been obliged to sell some areas to private investors. However, this free market economy cannot really succeed when the returns are so low; few individuals with access to the amount of money required are likely to sink large sums into such a poor investment. Accordingly, forestry has been made tax deductible, and most private forest buyers are using it as a way of avoiding tax on large sums of money. The situation is laid out very clearly in a briefing paper from the Ramblers Association:

> The golden rule for investors is to have a young plantation assessed under Schedule D and a mature plantation under

Schedule B. Under Schedule D a plantation is assessed under a normal profit and loss basis. Thus all the costs of maintaining a young, non-producing plantation appear on the tax return as a loss to be set off against a wealthy investor's other taxable income. Meanwhile the real value of the property is steadily increasing. But when a property changes hands it automatically reverts to Schedule B assessment unless and until the new owner opts for Schedule D. Under the Schedule B the income from sales of felled timber is not taxable, not even as capital gain. The only tax is a nominal one – one third of the annual rent value of the land as unimproved scrub. Thus the woodland owner who leaves his mature plantation to his heirs, or transfers it to a family trust, pays no tax while the asset is growing. His heirs or successors pay only nominal tax when they realize the value of the trees.*

In other words, forests are a convenient way for the rich to avoid state taxes. So-called 'private' forestry is still supported by the tax payer, but the nation will not gain the eventual benefit of the timber reserves. This inequality has been intensified by the recent Wildlife and Countryside Act, whereby landowners can be paid annually to leave ancient forest sites as they are, so that the profits from forestry can be paid for out of taxes without any timber being produced at all.

There are signs that these contradictions are being recognized in Britain. Two studies in the 1970s recommended a great increase in timber grown; these were *The Wood Production in Britain – A Review* by the Forestry Commission, published in 1977 and, even more extreme, the *Strategy for the UK Forest Industry* prepared by the Centre for Agricultural Strategy at Reading University and published in 1980. This latter publication called for the planting of an additional 1.8 million hectares by 2030, more than doubling what is there already. These proposals do not seem to have been taken up, and at the same time there has been a revival of interest in broadleaved woodland.

---

* Ramblers Association, London: *The Case Against Afforestation of the Uplands*, 1981

Many countries have avoided the extreme problems faced in the UK. Other central and northern European countries have a longer tradition of woodland management, so that a sustainable forestry practice is more likely to be continued. Nonetheless, there is controversy about the management of some central European forests at present; scientists believe that they may be exploited too intensively and that this could be a factor in current forest decline. There is likely to be an intensification in forestry practice in many countries over the next few years, and these could cause more problems similar to those experienced in Britain.

Afforestation in the tropics

The real debate about afforestation is likely to focus on the tropics over the next few years. There have recently been some enormous afforestation projects in the Third World, and while some of these have been quite successful, others have spectacularly failed to fulfil their original expectations.

There is an enormous difference between afforestation in tropical and temperate countries. In the tropics, rates of growth are far greater, and plantations can be mature and ready to use after 10 or 15 years, allowing a relatively rapid return on investment and a reduced need for long-term planning. Because future forest needs are inevitably going to rely increasingly on tropical plantations (as well as natural woodland management in the tropics) it is important to make correct choices about what to plant and how to plant it. Unfortunately, many of the mistakes made in temperate countries are being repeated in the tropics, cutting productivity and wasting land.

In Eastern Africa, native tree species were replaced by the more 'efficient' Mediterranean pine species, *Pinus radiata*. Initially, growth in highland areas was impressive, but later the trees were attacked by an indigenous fungus and large areas were destroyed. In similar way, eucalyptus introduced into India has suffered very badly from termite attack, although termites can live in balance with natural tree species. An enormous forestry project in the Brazilian Amazon, involving clearance and replanting of some 15,000 square kilometres (an area larger than Connecticut) by Daniel Ludwig, an eccentric American millionaire, was abandoned after only a few years, because of problems caused by climate and disease.

The Australian eucalyptus has proved a very mixed blessing to foresters. Popular because of its extremely high growth rate, it has a number of other problems, and is susceptible to native diseases. Rapid growth is accompanied by equally rapid water cycling; the soil can become totally parched in areas where eucalyptus is planted in quantity, damaging water tables and adding to agricultural problems. Overall, there is a strong argument for managing existing forestry wherever possible, rather than establishing new plantations, although it seems inevitable that more plantations will be established in the future.

Biomass and plantation forestry

One additional factor in the exploitation of plantation forestry is the rapid growth of interest in plant material ('biomass') as an energy source to replace fossil fuels. Biomass, like coal and oil, stores solar energy in a relatively compact form and can be converted into a number of solid, liquid or gaseous fuels. Although many of the current biomass technologies involve the exploitation of fast-growing herbaceous plants like sugar beet, various oil-containing plants, and even seaweeds, the use of trees is also important.

At its simplest, a forestry plantation, or even a naturally regenerating forest, can be used for firewood or charcoal, especially if the wasteful open fire is replaced by an efficient wood-stove. More recently three new or rediscovered echnologies aimed at speeding up the production of biomass have been discussed, and used experimentally. These are:

- Coppice plantations
- Single stem plantations of fast growing trees
- Whole tree harvesting, including use of roots, branches, leaves etc.

All of these have potential advantages and problems. Coppicing utilizes trees by carefully cutting them without actually felling, then allowing them to regrow, thus reducing the amount of ecological damage caused by clear-felling. Single stem plantations reduce growing time, while whole tree harvesting also uses branches, twigs and roots which are normally left in the forest. Of the three, coppicing is probably the most environmentally benign and is being promoted by conservationists in some areas to preserve sites as broadleaved woodland. Single stem plantations are usually monocultures, of little value to wildlife, and may also cause problems with nutrient depletion if cropping is repeated too often on one site. Whole tree harvesting greatly increases the amount of acidification resulting from forestry, to the extent that some countries have already abandoned it.

Biomass is already being used as a major energy source in terms of fuelwood, and it will continue to be used so long as people in poor countries have no other realistic option, however piously we object to the environmental implications. On the other hand, several developed countries are following a strategy of increasing their reliance on biomass, including Norway, Sweden and New Zealand. It is difficult to see how acceptable this option will eventually prove on a broad social and environmental scale, and it will have to be judged against the energy sources already established, including nuclear power and fossil fuels. However, it is certain that indiscriminate or badly managed biomass schemes will be harmful and any moves towards increased use of wood as fuel should be judged with a fair amount of scepticism.

4. Trees and disease

In its right and proper place, plant disease plays an important role in the evolution of species, eliminating weaker individuals and occasionally clearing whole areas during epidemics, so that new species can colonize. Unfortunately, modern forestry practices and the international movement of timber have increased both the incidence and severity of disease out of all proportion to that which is found in a natural ecosystem.

There are several reasons for this. Many pests and diseases of trees are specific to one or two tree species, so that in a natural mixed forest some filtering effect takes place; pests find it relatively difficult to move between host trees, and replacement tree species are available if one particular type of tree suffers badly for a while. Indeed, some forests probably have a natural cycle in which the dominant tree changes in a regular pattern over the centuries. In a single species stand, on the other hand, pests can spread extremely easily.

However, this cannot be the only reason for the increase in disease; many natural temperate forests are made up predominantly of one species, yet survive perfectly well. Plantation forestry also relies increasingly on imported species, which frequently bring their own pests along with them. Introduced pests, if they can survive at all, sometimes do very well for themselves, because few of their own predators will be present in their new home, giving them an advantage over native species. In Britain, for example, all conifers except for the Scots pine have been imported; many of their pest species were unknown here before the twentieth century. Even where a natural forest is managed rather than completely replaced, selective thinning of some trees can increase the concentration of others, making them more susceptible to pest attack.

It is not only imports of living trees that can cause problems. Today more timber is moved around the world than ever before, and imported wood can bring with it new pests and fungi, some of which will be capable of attacking native trees. Here the situation is even worse; not only are there no predators for the pests, but the native trees themselves have little natural resistance to the imported diseases.

Forestry also reduces tree resistance to diseases in more subtle ways, by steady genetic deterioration of timber stock. If an area is managed for any length of time, the best trees will be continually removed, allowing weaker individuals to survive to maturity and reproduce. Selective collecting of tree seeds and cones also sometimes acts against natural evolutionary development and further weakens the strain. The wounding of trees through timber extraction and as an accidental result of tourism or sport gives fungi a greater chance to establish themselves. The role of tree damage caused by humans in spreading disease is especially important in many tropical forests, where the humid conditions make fungal attack especially likely.

In large parts of Europe and North America, the spread of tree diseases is being further affected by human actions of a less direct kind, through the mounting levels of air pollution found over very large areas. As the following chapter will show in more detail, sulphur dioxide, nitrogen oxides and ozone combine with other pollutants to damage trees. One of their most significant effects is probably to increase trees' susceptibility to various diseases.

Where diseases do strike, the effect can be rapid and devastating. In Europe, one of the most disastrous examples in the last few decades has been the spread of Dutch Elm Disease and the virtual eradication of the elm from many countries, as a result of a disease which was almost certainly imported.

Dutch Elm Disease

For many people living in southern Britain, the English elm was part of the very essence of English heritage; tall, stately and producing excellent timber, the elm dominated large areas of lowland countryside for hundreds of years and was celebrated by artists as diverse as Constable and Turner.

The landscapes that these artists painted have now disappeared forever. Between the late 1960s and 1977 it is estimated that well over 10 million elms died in Britain, victims of the 'Dutch Elm Disease' which has virtually eradicated elm populations throughout huge areas of Europe and North America, and which has made tree disease an issue of immediate relevance to governments and individuals throughout the developed world.

Dutch Elm Disease isn't really Dutch at all. Its effects were discovered in France in 1818 and finally identified as being the work of a fungus by tree specialists in the Netherlands, almost exactly 100 years later. Although no one has ever proved where it originated from, most specialists accept that it is of Asian origin and that it entered Europe with imported timber. Certainly, its sudden spread suggests either a new mutation or an introduced pest.

The fungus which causes all the damage is similar to a yeast and is called *Ceratocystis ulmi*. It lives in the sap vessels of the elm and releases toxic substances which stimulate the vessel walls to produce bulbous, gum-filled enlargements called tyloses; these stop the flow of nutrients and gradually starve the tree's extremities of food. First signs of damage include drooping and incurled leaves, usually in just one small area of the crown. As the fungus multiplies, it gradually spreads outwards, affecting a large area and usually killing the tree completely in two or three years. The inexorable spread of the disease is reminiscent of cancer and may be one reason why Dutch Elm Disease is regarded as so sinister.

Although the disease can spread within a tree simply through the reproduction of the fungus, its movement between trees is largely controlled by a carrier, the bark beetle *Scotylus scotylus*. These beetles breed beneath the bark of recently dead elms, where the larva burrow long tunnels into the surface of the wood, producing a characteristic pattern. Once mature, the beetles fly to living elms with some of the deadly fungus attached to their bodies, and 'inject' it by puncturing the bark of young branches to feed on the tree sap.

Dutch Elm Disease was first noticed in Britain as long ago as 1927. It was widespread by 1928 and continued to flourish until 1937, whereupon it began to decline and appeared to have 'burnt itself out'. By this time, arboriculturists estimated that between

10–20 per cent of British elms were already dead. However, a more virulent form appeared in the USA during the 1960s and started to kill North American elms at a rate of some 400,000 a year. It reached Britain in 1970 and, despite a concerted campaign by certain local authorities, had killed 4–5 million within four years. A total of 11 million was reached by 1977, and now most of the 20 million alive before the epidemic have disappeared. Although trees are still killed, the vast majority of elms in the UK have already died, leaving many hedgerows and woods virtually empty of trees.

What is so frightening about Dutch Elm Disease is the apparent impossibility of doing anything about it once it has appeared. 'Cures' have consisted of either felling dead trees and removing the bark, thus stopping the beetles from breeding, or undertaking time-consuming medication by injecting fungicide into the trunk. In practice, the latter process was far too slow for any but a few prize trees; only about two or three are treated per day and the chance of treatment being successful is small, in any case.

The fact that imported timber played an important part in spreading the disease now seems indisputable. High concentrations of disease were seen in the Severn-Avon valleys, south Hampshire, south-west Sussex and Essex, all near major ports where the more virulent strain could be carried into the country with timber supplies from abroad. The fact that the fungus was imported means that it had no natural enemies in Britain and that the elms were not able to build up adequate defences.

In some cases, natural changes also cause a sudden increase in disease. When a great storm blew down part of a Colorado spruce forest, most of the standing dead trees were flattened. This reduced the breeding places available for woodpeckers, which usually peck out nest holes in dead timber, and the birds therefore had a sudden population decline. This meant that one of their food insects, a bark beetle which had previously been beneficial in thinning out senile trees, increased rapidly over three years and caused a similar spread in a fungus. Once established, the disease caused the death of millions of trees over the next six years.

The spread of Dutch Elm Disease has been carefully recorded and there is good evidence that its effect has been partly or wholly of our own making. Devastating tree diseases are not impossible in natural conditions, and disease is an important evolutionary factor. But at present we are greatly increasing the chances that these diseases will become more common and will spread further than would otherwise be the case.

Other British trees are under threat as well. The native English oak is failing to reproduce successfully, due to attack by a small wasp which arrived from the continent in the early 1960s. Strangely, these parasitic wasps form galls on both the native English oak and the imported Turkey oak, but only the acorns of the native tree are damaged. Scientists fear that, as the old trees die, they will be replaced by the Turkey oak, which is far less valuable from a wildlife point of view. Oaks could be in danger more quickly, however. Oak wilt, which is similar to Dutch Elm Disease, is already endemic in North America. While timber imports are strictly controlled, illegal smuggling of timber is already common and scientists believe that it is only a matter of time before the disease arrives in the UK. If this does happen, the whole ecology of many rural regions could be substantially damaged and vast woodland areas destroyed.

Ash trees are also under attack, this time by a dieback disease. A 1983 survey showed that half the British trees over 25 years old between Humberside and Buckinghamshire were suffering dieback. As one arboriculturist has said, 'All of a sudden we will wake up and say "Good God, where are all the trees?".'

Tree diseases throughout the world

The USA has suffered particularly badly from tree disease. An estimated 15 billion elm trees were lost there in the course of the Dutch Elm Disease epidemic, while an earlier disease virtually wiped out the American chestnut, which once made up 25 per cent of the trees in the eastern states. Even in strictly financial terms, these diseases are disastrous, and it has been calculated that 30 billion board feet of chestnut were lost because of the disease.

Many of these outbreaks have been traced back directly to human interference. A beech scale insect, introduced to Nova Scotia in 1890 along with ornamental beech trees, acts as a carrier for a fungus which attacks the bark of beech and has destroyed large forests. Selective harvesting has increased the concentration of beech in many areas; because it seasons badly and is often left by loggers, the disease has even more chance to spread. Another harmful insect, the gypsy moth, was introduced deliberately when it was used in an unsuccessful attempt to breed a hardy silkworm; now it has spread rampantly and defoliates oak, making the trees more susceptible to fungal or insect attack.

Unfortunately, not all diseases are even specific to a single type of tree. A soil fungus, *Phytophthora cinnamoni*, which is closely related to the infamous potato blight, is known to have attacked over 400 tree species of 48 orders. Since appearing in Sumatra in 1922, it has spread throughout much of the world and is particularly prevalent in Australia and New Zealand. It is thought to be spread either as wind-blown spores, or perhaps also on car tyres, as outbreaks seem commonest beside forestry tracks. Hundreds of thousands of hectares of eucalyptus have been lost in Australia, including particularly the jarrah tree which is important for watershed protection.

At the present time, the largest outbreaks of disease have been in the more industrialized countries, where human interference in forestry practice is greatest. As more plantations are established in the tropical areas and the diversity of trees is reduced, we can expect to see a similar increase in tree diseases there as well.

Spraying against tree diseases

Tree diseases do not only mean losses of habitat and timber. For people working in forests, or living nearby, the attempts to control forest diseases provide a new health risk from insecticides, herbicides and fungicides used to keep down any unwanted animals and plants. Sprays are applied from hand-held backpack sprayers, from trucks and from aeroplanes and helicopters. Drifting spray threatens the workers applying it directly, and also blows across neighbouring land, harming livestock and people.

In the last 20 years, the application of pesticides has increased out of all recognition. Herbicides are used routinely to keep down 'weed' species (i.e. virtually anything apart from the trees themselves), and insecticides are applied against the growing numbers of pests which are utilizing the rich pickings of the coniferous monocultures.

All application methods have their own dangers. For the operator, the backpack is probably the most dangerous, because the sprayer will routinely have to direct the spray in front of his or her body, so that it will blow directly into the face. For people living in or near forest plantations, aeroplanes or helicopters are the worst option, because they spread the chemical far further and thus make it more likely to drift. Recent research has shown that a great deal more of the chemical is released in very small droplets than was previously thought, so that a greater proportion is liable to drift uncontrollably in the wind.

Some of these chemicals are undoubtedly dangerous. Even in the developed countries, 2,4,5-T is still widely used, despite its clear links with cancer and the mutations that arose from its deployment as the infamous 'Agent Orange' in Vietnam. Cases of birth defects and cancer continue to build up, and many countries have banned the use of 2,4,5-T; others, including Britain, claim that current formulations are safe. Other permitted chemicals are also highly toxic. Sprayers are instructed to wear special clothes and to avoid handling any liquid, yet these same sprays are indiscriminately released into the air. In many cases, we do not know the effects of long-term exposure to small doses, even though the acute toxic effects have been measured.

For foresters in the Third World, the situation is much worse. Many chemical companies 'unload' pesticides, banned in the richer countries for safety reasons, onto poorer countries which have little pollution or industrial safety legislation. To make matters worse, many workers in the under-developed nations are not taught how to use the deadly chemicals safely, further adding to the risks they incur.

Current timber and forestry practices have made the need for disease control far greater than would be required in a more natural forest. The push for profits has increased pressures on plantation managers to use sprays more frequently. At the same time, the chemical companies have used their enormous economic weight to promote further chemical use. For many people living and working in rural areas, the cure for tree diseases may be more deadly than the disease itself.

However, it is not at all clear that spraying necessarily even does much to reduce disease. Pests develop resistance to sprays very quickly, and many of the pesticides used also kill predators of the pests, so that the surviving pests can breed even faster than before. A comparison of forests in California which were sprayed against Douglas fir tussock moth with those which were left untouched, found very little difference in total tree mortality. Spray programmes against the gypsy moth and the spruce budworm have been similarly unsuccessful.

From a conventional forestry point of view, sprays are an essential part of management. However, tree diseases themselves are largely a result of mismanagement on a broader scale, and the resultant chemical applications are no more than a rather dangerous type of first aid, comparable to taking drugs against bronchitis whilst continuing to smoke tobacco. As such, an examination of tree diseases cannot be separated from a more comprehensive examination of forest policy.

5. Trees and pollution

By far the most significant development in Europe and North America over the past five years or so has been the sudden realization that huge forest areas are being badly affected by pollution, including sulphur and nitrogen oxides, ozone and 'acid rain'. Forest damage was first noticed during the late 1970s in parts of Germany, including the Black Forest, but it is only recently that the role of pollution has been widely acknowledged by foresters. The precise mechanisms involved in tree death are still uncertain today.

The scale of the damage

In Europe, the scale of damage is already quite remarkable. Some 34 per cent of the Black Forest in Germany, about 2.5 million hectares, is now showing signs of serious tree decline, including crown thinning, premature senescence and death of trees. Eighty per cent of the country's firs are already affected. Elsewhere, 500,000 hectares are damaged in Poland, another half million hectares in Yugoslavia, and in Czechoslovakia scientists are talking about 60 per cent of the forest areas showing signs of serious stress – 40,000 hectares in the Erz mountain range are already dead. The Soviet Union is also suffering tree death, and although official figures are not often available, a recent article in *Pravda* revealed that vast forest areas are dying near industrial areas 1,300 kilometres east of Moscow. Switzerland, Austria, Belgium, Romania and East Germany all have similar problems.

The extent of the damage has seriously frightened many Central European countries, where forests play an important role, not only in economic terms, but also as a major amenity resource, and as an integral part of the folk culture and identity of the society.

Opinion polls show that, for the Germans, the threat to their beloved Black Forest is one of the most crucial public issues. The German government has been spurred into making serious attempts to cut down pollution and to halt the destruction.

Alarmingly, effects are now appearing in less polluted regions away from the heavily industrialized Central European countries. These have recently been reported in the Netherlands, France, Finland, Sweden and Norway, despite the Scandinavian countries believing, until recently, that they had escaped damage. Conifer plantations in the British Pennines were abandoned from the 1930s onwards because of pollution effects from the northern cities. Whilst these acute toxic effects have now been reduced, pollution damage in Scandinavia is at the same pollutant level as that still found in the UK. A German forester visiting Britain in 1984 found signs of damage identical to those in the Black Forest, although some British experts believe that these are due to climatic factors rather than to pollution.

In North America, substantial dieback is taking place in the Adirondack National Park region of New York State and in the Sierra Nevada region of California, where freshwater acidification through acid rain has already occurred on a large scale. Although damage is nowhere near as severe as in Europe, researchers recently concluded that no other events in the trees' growth history were 'as widespread, long-lasting and severe in their effects' as those caused by air pollution.

Implications of forest damage

The ecological implications of widespread forest death are extremely serious. Wildlife suffers, soil will be subject to increased erosion and large areas of important natural habitat will be lost, perhaps forever. However, there are also clear and important consequences, connected with pollution damage, for the economy and for employment. In countries with large forest areas many people are employed in managing and utilizing the forest; the products form both a valuable internal commodity and an important export.

In the Federal Republic of Germany, conservative estimates are that one billion dollars' worth of timber has already been destroyed by pollution; if present trends continue, this figure will be increased many times over. In addition, the price of preventing pollution (which often has to be paid by people living in a different region or country) is already very considerable, including costs for liming soil, extracting dead timber and researching into effects and remedies. Similar costs are likely to affect all the Central European forest industries and may be significant in Scandinavia, France and Britain as well.

According to Dr Raymound Brouzes of Environment Canada, the official Canadian environment protection agency,

> A reduction in tree productivity as small as one per cent or even 0.25 per cent per year would result in a significant reduction in total wood production if compounded over the lifespan of a tree. Such a reduction could have serious effects on the fibre supply and the economic well-being of forest-based industries.*

Current European losses are already considerably greater than this.

Causes of forest damage

The precise reason for the dieback has been the cause of intense argument throughout the developed countries. No single factor appears capable of explaining every case of forest damage, and it now seems likely that the overall pollution load combines with environmental variables like climate and geography to weaken trees, making them more susceptible to disease or killing them outright.

The pollutants sulphur dioxide, nitrogen oxides and ozone are all involved to some extent, both through their direct effects on living trees and probably also because they are changing the soil habitat through acidification, killing micro-organisms, depressing the decomposition rate and so on. Although some sulphur and nitrogen oxides will occur naturally, coming from volcanic

---

* Sandra Postel, 'Air Pollution, Acid Rain and the Future of Forests', Worldwatch Paper, 1984.

eruptions, ocean sources and biological processes; most of those found in the atmosphere of Europe originate from artificial sources. Sulphur dioxide comes mainly from burning fossil fuels in power stations, factories and domestic houses, while the largest source of nitrogen oxides is the motor car, although any combustion process releases some nitrogen oxides.

Once in the atmosphere, these gases can affect trees either by landing on them in the form in which they were released, or by being incorporated into clouds and falling as 'acid rain'. In addition, nitrogen oxides are major sources of ozone, formed during hot, sunny weather, and known to damage trees and crops in the USA.

However, it is clear that pollutant effects are increased by environmental factors as well. In West Germany, the first trees to show signs of stress are often found at high altitudes, where they will experience greater extremes of temperature; damage is commonest at the edges of forests in places where trees receive the most wind and rain. In addition, high forests are subject to far more cloud and mist and some scientists believe that acids reaching plants in clinging mist (so-called 'occult' precipitation) are more damaging than those found in rain.

Drought also seems to be a factor in forest death, and some timber specialists have been tempted to attribute all the Black Forest damage to climatic effects. This hypothesis is no longer accepted by many experts; although drought-induced dieback has occurred in the past, nothing on the present scale has been seen before, either in the extent of damage or the range of species involved. In addition, it is now known that forests were suffering dieback in Germany before the great drought of 1976, which has often been quoted as a possible initial cause.

The very complexity of forest damage has led to a confusion among investigators, and a proliferation of 'comprehensive' theories which do not necessarily stand the test of time. Ozone damage has been a popular theory because it is known to have killed trees downwind of Los Angeles in California. Ozone levels have apparently changed far more dramatically in Europe in the last few decades than concentrations of sulphur dioxide. However, ozone alone cannot explain all the damage; dieback is also found in places where ozone concentrations are not particularly significant. It is possible that both ozone and sulphur dioxide are involved, and that their combined effects are greater than the sum of their individual effects, perhaps through ozone damaging the cuticle of leaves, allowing sulphur to attack. This 'synergistic' effect (i.e. an increased effect of two or more pollutants), between ozone and sulphur dioxide, is already known to take place for some crop species.

Another theory is that the acids in rain have gradually acidified the soil, reducing the rate of nutrient cycling by killing soil micro-organisms and releasing toxic metals, including aluminium, which damages root hairs and further weakens the trees. This theory, proposed by Dr Bernard Ulrich and his colleagues working in the Solling district of the Federal Republic of Germany, has been quite heavily criticized recently by scientists who believe that direct toxic effects are more likely, but it has yet to be disproved. If true, or even partially true, it is extremely serious, because acidification of the soil will not be so easy to reverse as airborne concentrations of pollutants.

It is unlikely that any final decisions about pollutant pathways will be reached very quickly, if at all. Ecosystems are too complex for simple explanations, and we can expect rapidly changing public and official attitudes over the next few years, depending on the latest estimates of damage or recovery. (If drought really is a significant causal factor, along with pollutants, then we may well see periods of 'recovery' interspersed with renewed damage, and at each change fresh arguments about the necessity of pollution control are to be expected.) What is becoming increasingly clear is that air pollutants are involved in the process of tree deaths in some places, and that a reduction of damage depends, ultimately, on a reduction in pollution.

Sources of pollution

To some extent the theory that a number of pollutants are involved in forest death will, if it is proved correct, be a vindication of early environmental fears about the cumulative effects of discharging pollutants into the atmosphere and hoping that they will simply go away. Air pollution load has gradually been increasing in Europe and North America, although some individual countries have taken steps to reduce certain pollutants, including Britain with its various Clean Air Acts. Several factors have combined to make the current situation especially severe. Perhaps the most significant of these will prove to be the great increase in nitrogen oxides relative to sulphur dioxide. While sulphur emissions have stabilized or fallen, due mainly to the levelling out of energy demand and the economic recession, the amount of nitrogen oxides has increased dramatically in the last few decades, due to the increasing number of motor vehicles in use. There is no indication that this process is slowing down, so further increases are likely in the future. Thus, while nitrogen oxides currently comprise only about 30 per cent of total acidic emissions, their proportion is continuing to increase. This is important, both because the total pollution load increases, and also because a mixture of pollutants is often more dangerous than a single pollutant.

At present, damage is almost entirely confined to the developed north, except for very localized effects around large cities in some Third World countries. However, acid rain has now been recorded from increasingly remote regions, including China, rural Yugoslavia and the Amazon. Tropical soils are believed to be particularly susceptible to acidification, so that any increase in air pollution levels within tropical countries should be regarded as particularly disquieting.

While pollution usually comes from a wide variety of sources, the majority of pollutants are usually from a few large sources, like power stations or factories. Indeed, one huge smelter at Sudbury, Canada, has the dubious honour of producing 1 per cent of the world's total sulphur dioxide all by itself. Thus any control strategies are most logically aimed at the main polluters. Unfortunately, these polluters often wield a proportionately larger political clout as well, and have frequently resisted attempts to install expensive pollution control equipment. Governments in Europe and North America are beginning to realize just how difficult enforcing pollution controls is likely to prove in practice.

6. The burning forests

> On examining those sections whose trees are a hundred or
> two hundred years old, we find the same fire records,
> showing that a century or two ago the forests that stood
> there had been swept away in some tremendous fire at a
> time when rare conditions of drought made their burning
> possible. (John Muir, Olympic Mountains USA, 1918)

Fire is as much part of many forest ecosystems as other natural
disasters such as avalanches, periodic windstorms and outbreaks
of disease. Plants and animals in such areas have adapted to fire,
and some even appear to require it for their long term survival. The
lodgepole pine of northern USA produces many cones which will
only open and release their seeds after exposure to intense heat.
The Douglas fir, the dominant tree in the same area, has adapted
to sporadic fire outbreaks by evolving thick, insulating bark and
seeds which germinate best in open conditions and in mineral soil
such as is formed after a forest fire.

In some areas, a build-up of dead timber is an indication that
fire is likely, clearing away old trees and providing space for
seedlings to develop. Fire also reduces the chances of any one tree
species taking over the forest completely and crowding out all the
others. Forest rangers in the Yellowstone Park in the USA predict
that the area will become susceptible to a major fire from about
2050 onwards because of a build-up of dead timber, but that until
then any accidental fires are unlikely to spread very far.

Unfortunately, forest fires are no longer just a product of
random droughts and lightning storms. Humans have increased
the odds out of all proportion to natural causes. The growth of
forest tourist industries has increased the risks, as has the
popularity of camping, often now practised by city dwellers with

little knowledge about basic woodcraft or of how fast a fire can spread. Accidental fire is now one of the major hazards facing foresters producing timber in heavily populated areas.

Modern forestry practice has contributed its fair share to forest risk as well. Planting trees artificially close together and draining marshy land both increase the chances of fire spreading. The practice of leaving gaps ('fire breaks') to stop fire spreading has been shown up as inadequate on dozens of occasions. In Britain, the vast majority of forest fires occur in conifer plantations, even though the native broadleaved woodlands are more heavily visited by tourists and walkers.

Forest fires are an additional danger where the ecosystem is particularly fragile, or where humidity is low so that fires are likely to spread once they start. The taiga in northern USSR is exceptionally prone to fire from campfires and other accidents, especially in the ecologically important northern Baikal region. Hundreds of thousands of hectares are apparently destroyed annually according to reports from within the Soviet Union.

Over the last few years the threats to trees have been increased by growing evidence of fires started deliberately. While a proportion of these are due to vandalism, other fire raisers are working to a deliberate policy. Vast fires have been started all over the Amazon as people clear forest for the plantations and ranching. But fire raising is by no means confined to the Third World. In Europe as well, intentionally started fires have destroyed tens of thousands of hectares of prime forest land.

There are several motives behind burning timber. Insurance claims have been the reason for fires started in the south of France. Attempts at intimidation of people owning land required by another organization are another reason. Scores of forest fires in southern Spain in 1984 were blamed on criminals working for sections of the timber industry, because scorched wood can be bought more cheaply than undamaged timber but is still fit for many uses. In Britain, woodland and hedgerow fires are frequently started as a side effect of deliberate stubble burning.

Forest fires today are seldom just the result of natural ecological processes. Fire raising, whether accidental, as a criminal act, or as 'management policy', is now an additional threat to forest survival in many areas.

7. A global overview of forest prospects

The hills of Lebanon were famous for their trees for thousands of years. They are mentioned specifically in the 'Epic of Gilgamesh' which may be the oldest recorded story in human history. The most famous trees were, of course, the cedars, but pine, fir, juniper and oak all grew as well, providing abundant sources of timber and a rich environment for the people who settled there. Today, there are just 12 small groves left, protected by their inaccesibility or their proximity to sacred shrines. The hills are bare desert or scrub and most of the people living there today probably believe this to be the natural state of their land.

The important point here is not that some unscrupulous multinational company has exploited the forests in the last 20 years, or that expanding populations have been forced to use the remaining trees for timber. The deforestation of the Lebanon began about 5,000 years ago and was effectively finished centuries before the modern interest in ecology.

To the best of our knowledge, the first serious exploiters of Lebanon timber were the Phoenicians, the entrepreneurial traders who bartered their way all over the Middle East and exported timber to the Egyptian Pharaohs. These early businessmen were later copied by none other than King Solomon, who felled huge quantities of Lebanon timber to build his temple, if biblical accounts are to be believed. However, many trees must still have survived at this stage, because centuries later Alexander the Great of Greece cut down more to build his Euphrates fleet. By the time the area was annexed by the Roman empire, the once beautiful forests were in ruins. The emperor Hadrian tried to conserve the remainder by laying down laws to this effect. Perhaps he was already too late, for the forests were by then a fraction of their original size, and when Lebanon was abandoned by Rome, settlers quickly moved in with goats which destroyed a large part of what remained, while the rest was felled by firewood traders.

Lebanon's forests were destroyed for archetypal reasons: large buildings, celebrating dictators, the machinery of war, trading in energy, and bad agricultural practice. Of course there may have been many other reasons which have escaped the scanty history of the period, but two facts stand out; first that deforestation is by no means a new phenomenon and second, the effects of deforestation can be permanent in some situations.

The latter point cannot be emphasized enough, especially to a western reader who is familiar with regular replanting schemes in temperate regions; however ugly and ill-conceived the schemes may be, at least the trees usually grow! In tropical or sub-tropical places, deforestation can take place on soils too poor to survive long without a protective tree cover, and loss of forest can alter climate so that conditions no longer exist for regrowth. Even in the twentieth century we can do little about these old disasters; the energy and water resources to reverse the effects are simply not available.

The Lebanon is not an isolated example. In Africa, the third of the country that is now 'natural' savannah was forested before early settlers burnt and cleared huge areas. Further north, in Morocco, Algeria and Tunisia, forests used to cover 40 per cent of North Africa but are now confined to less than 10 per cent, much of it plantations of imported species. It is now thought that the famous Easter Islands in the Pacific were actually abandoned not because of war or sickness, but because severe deforestation made living conditions intolerable. Many of our areas of desert and scrubland were once lush, productive forests, but it is difficult to see how they ever will be again.

Despite the horrifying statistics, it is very difficult to get an overall picture of what is happening to the world's forests. At worst, pessimists may be tempted to conclude that things have already progressed so far that there is little point in doing anything. At the other extreme, large corporate organizations are still acting as if timber were an everlasting resource. (There are sound pragmatic reasons for them to do so and these will be discussed later.) In the following pages an attempt is made to give a general picture of the world's forests and timber stocks.

Temperate forest

As the previous chapters have shown, the problems of temperate forests are not so much the decrease in overall forest area, but the substantial change in the types of forests and their ecological diversity and stability. In very broad terms, the total area of forest in Europe, the USSR and North America is likely to decrease only slightly in the medium future, but the proportion of forest existing as plantation is liable to rise sharply. This will lead to an increasing conflict with recreation and conservation interests and will also tend to make forestry a more capital-intensive operation, hence providing fewer jobs.

Europe
Apart from the far north of Scandinavia and parts of Russia, most European forest areas have a long history of human interference. Many present-day forests have, at some time in their past, been cleared, a factor which has important ecological consequences. European forest land can be divided into three main types:

- Forests that have survived essentially unchanged by human interference. These are confined to comparatively large areas of Scandinavia and the USSR, and small fragments in other countries.
- Forests that have been destroyed at one time but have now either regenerated naturally or been deliberetly replanted and managed. These include most British forests, large areas of Central Europe and southern Scandinavia.
- Areas that were once forested but are now either agricultural land, urban areas or scrub and semi-desert. Cleared areas include much of Britain and northern Europe and the scrubland in the Mediterranean and Aegean regions.

British forest
Britain provides a good example of what can happen to forests in the more populated regions of Europe. For much of the 10,000 years after the Great Ice Age, virtually all of the British Isles (including the 'natural moorland' areas of upland Scotland

and Wales) was covered with extensive forest. First, birch and Scots pine covered most of the area, with willows predominating in wet lowland areas. After 5000 BC the weather became wetter, and initially oak, elm, lime and alder started to dominate the dry land, followed later by ash, beech and hornbeam. The trees were probably at a very low density by today's standards, with diverse ground flora, and were the home of deer, wild boar, bears and wolves as well as the smaller mammals and birds which are still found in Britain.

In the uplands, more birch and alder were found, and large areas of Wales may have had a very slow rotation of species between oak and alder. Further north, in Scotland, vast areas of Caledonian pine forest consisted of very open Scots pine forest. Because of the isolation of Britain during the latter Ice Ages, many tree species found on mainland Europe were absent. Scots pine is the only conifer which is naturally present – all the others have been introduced. Even common broadleaved species like the sycamore, horse chestnut and sweet chestnut were only introduced at the time of the Roman invasions.

These forests survived unchanged for many thousands of years. Early human settlers simply lived among them, in much the same way as the aboriginal peoples of today's rainforests do. However, a major change occurred during the time of the 'Neolithic Revolution', when the population expanded very rapidly. During this time, many of the natural forests were cleared in Britain, a situation that was perhaps also intensified by the effects of a slightly drier climate, which was less suitable for some tree species. By the Iron Age much of southern England was under the plough. (Here, 'rapidly' should be taken very relatively; compared to the rate of forest destruction seen today the expansion was very slow, taking many centuries, but was quick in terms of the historical development up to that time.)

Nonetheless, Julius Caesar still described Britain as a forested country, and found that resistance fighters could use the extensive woodland as cover. The next main clearances came with the Saxons, who brought their livestock with them and converted huge areas into pasture. By the time the Domesday book was written in 1086, deforestation was well advanced, with 80 per cent of the original forest already gone. However, there were still extensive forests left well into the Middle Ages and beyond, like the famous Sherwood Forest in the north and the Midlands, and the Caledonian forest areas of eastern Scotland, although changing climate had already spelt the end of many western Scottish forests. These early 'nature reserves' were jealously guarded, not out of any respect for ecology, but because the feudal overlords used them for hunting. Deer stocks were built up, especially as the larger 'game' such as bears and wolves were hunted to extinction. The native red deer were the most favoured species, although fallow deer were also introduced from Europe. Today British red deer are noticeably smaller than their continental cousins, perhaps because selective hunting of the strongest specimens has gradually weakened the race.

However, the remaining forests outside the park did not survive the next few decades at all well. The peasants were given 'common rights' to the land, and proceeded to graze pigs which ate the nuts, thus stopping woodland regeneration. Firewood was also collected on an increasing scale so that whole areas were gradually destroyed. In the Middle Ages, the Cistercian monks cleared more for their huge sheep farms in many of the remaining forests. By the fourteenth century, Britain was already importing timber.

The 60 or so 'royal parks' remained relatively unscathed for several hundred years. Their eventual demise came when the power of the feudal lords was on the wane and an even stronger incentive emerged: the need for timber to produce weapons of war. During the following centuries, large areas were cleared to build the powerful ships of the British Navy, effectively destroying much of the remaining forests in Britain. A new 'first rater' ship for the Royal Navy required no fewer than 4,000 large trees (mainly oaks), equivalent to felling more than 100 acres of forest.

The beginnings of forest management gradually developed. Coppicing trees (i.e. cutting them so that they regrow and can provide a regular crop) and 'pollarding' the tops (to produce a similar effect) both became popular. Despite the increased efforts, the demand for timber to smelt iron ore was so great that Elizabeth I banned smelting in Sussex and Furness to save the forests, and the industry moved to the Caledonian forests in Scotland.

Stocks had fallen to an all-time low at the time of the Great Plague. The Fire of London in the following year, as well as the sinking of the British fleet by the Dutch, created an unprecedented demand, and there was insufficient oak for the task. In the aftermath, a politician named John Evelyn lobbied for greater efforts at planting, resulting in an Act of Parliament in 1668 ordering extensive afforestation. The timber established in the following years was eventually used to build Nelson's fleet in the Napoleonic Wars.

The next major crisis came during the Industrial Revolution. Charcoal was needed in quantity, and charcoal makers became the new exploiters, building temporary encampments in forest areas and making charcoal on the spot. Their inefficient methods wasted far more timber than necessary but the job shortages created by the Enclosures Act, and industrial growth, gave little choice to people remaining in rural areas. If the discovery of coke had not been made, our few remaining woodlands would probably also have disappeared. Rising populations put additional pressures on forests for fuelwood and construction materials.

By the eighteenth and nineteenth centuries, British forests were in an extremely poor state. At that time, many independent landowners took decisive steps towards reforestation, employing cheap labour to plant up forests once again. However, there was less enthusiasm in the second half of the nineteenth century and planting again fell off, until the experience of two world wars emphasized the political requirements for strategic reserves.

Throughout British history, the requirement for weapons has been a decisive factor in the destruction of forests and depletion of timber stocks, much as it was for the ill-fated cedars of the Lebanon. After the First World War was over, the new Conservative government created the Forestry Commission in 1919, with the specific aim of building up timber stocks against the advent of another war and the renewed need for large supplies. Charged with the task of doing their job as cheaply as possible, the Commission chose conifers as a fast-growing alternative to traditional British broadleaved trees and pioneered the rigid plantation policy of straight lines and single species which now dominates so much of Britain's upland areas.

Since 1945 we have seen, once again, a large increase in the pressure on natural or semi-natural woodlands. Ecologists estimate that half the ancient woodlands in existence at the end of the

Second World War have been destroyed; a very significant statistic in view of the already impoverished nature of British woodlands. Destruction has been caused mainly by replacement with conifers (a more immediately marketable 'crop'), clearance of areas for agriculture and for fuel. At the same time the hedges, which have provided a valuable alternative to mature woodland for many animals and plants, are also being ripped up at an accelerating rate to make way for larger fields.

The defence of Britain's remaining woodland areas is already a contentious and vigorously contested issue, with land-owners claiming that middle-class urbanites are interfering with their rights and conservationists protesting that irreplaceable woods are being destroyed for short-term gain. The recent Wildlife and Countryside Act has probably done more harm than good in the long run, because owners can now claim large sums in compensation for not damaging woodlands and other important sites, yet the money to pay all these claims is not forthcoming from the government.

There are other threats to British trees as well. In many areas of upland Britain, sheep are under-grazing woods, so that no new saplings can establish themselves. The result is that many of the Welsh woodlands, so beloved by the tourist industry, are slowly dying; something like 93 per cent of those in the Snowdonia National Park are believed to have a lifespan of less than 100 years unless steps are taken to exclude sheep and replant oak. Even Britain, with a more conservation-minded public than in most other countries, faces the real possibility that natural woodlands will soon be confined to a few nature reserves.

Mainland Europe

In Mediterranean regions much of the original forest has been destroyed long ago and, because of the poorer soils, it has never regrown. The lands of the Yugoslavian coast and islands bear stark witness to the glories of Venice; 300 miles of coastal forests were plundered by the Venetians to shore up their great city. Elsewhere in the Mediterranean, deforestation has taken place more gradually, due to a steadily rising population and the spread of sheep and goat farming.

Nonetheless, forests are one of Europe's few major natural

resources, providing that the current disastrous spread of pollution damage does not continue indefinitely. Europe has 155 million hectares of commercially exploitable forest, with another 9 million hectares of forest which is unexploitable either because it is inaccessible or because it is already designated as nature reserves. Afforestation over the last 20 years has been at a rate of about 150,000 hectares per year and this is expected to continue. There has already been considerable re-afforestation in Nordic and Central European countries over the last few centuries.

This forest wealth falls largely in the hands of the Scandinavian countries. The ten EEC countries have just one per cent of the world's forests, although they account for 30 per cent of the timber trade. The EEC has 0.13 hectares of forest per person on average, as opposed to 2.4 hectares per person in the USA. But for most countries it is far less than that, as 85 per cent of the total is concentrated in France, West Germany and Italy. There is no EEC forestry policy as such, although there have been a few directives about controlling tree diseases and about the genetic quality of reproductive material. A policy may be developed in the future. Over half the forest area is privately owned and public access is limited to about half the total area.

Distribution of forests in the European Economic Community

| | | Forest Area | |
	total (1,000 ha)	as % land area	ha/head pop
Belgium	615	20	0.06
Denmark	470	11	0.09
West Germany	7,200	29	0.12
France	13,950	25	0.28
Ireland	330	4	0.09
Italy	6,300	21	0.12
Luxembourg	85	32	0.24
Netherlands	310	8	0.02
United Kingdom	2,020	8	0.04
Greece	2,500	19	0.26
EEC	33,780		

Source: Food and Agriculture Organization of the United Nations, 1983

The major change likely in the next few decades is an intensification of management. Currently, much of the regrowth is still due to natural regeneration; increased timber requirements may well encourage this to be replaced by more active planting and consequently this will create less diverse forests. *The Global 2000 Report to President Carter*, which provides an admirable summary of the world's forest prospects, comments:

> As management intensifies forests will become younger and still less diverse. This will lead to a reduction of some ecological niches and is likely to cause the extinction of some plant and animal species and changes in the population dynamics of others.

However, the process of intensification is not likely to go unchallenged. Over much of Central Europe there is a strong tradition of forest management and, more importantly, the forests play an important cultural and social role in many countries, so that there is always strong public pressure against destruction. This interest has been built up through hard experience. A few hundred years ago, deforestation in the Alps was so severe that soil was eroding rapidly in many places and there was a firewood crisis. Concerted efforts to replant have paid off and the large forests seen today, whilst technically often 'artificial', are a sustainable source of timber and an invaluable reserve for wildlife, in addition to maintaining watersheds and soil.

The intensification is proposed because it is calculated that timber requirements will rise by 45-80 per cent by the year 2000. Such sweeping generalizations about increases in consumption of raw materials have proved incorrect in the past and these figures are based largely on an increasing demand for paper and wood-based panels. The role of recycling, political and social choices about packaging, and general economic factors could all combine to reduce net requirements. Future trends will depend at least partly on how effectively the conservation lobby can publicize the links between consumption and forest degradation. At the same time, more conventional foresters are becoming disenchanted with some intensive forestry practices for all sorts of practical reasons and this may change attitudes to forest management in the future.

Whatever happens, it is almost certain that the largest bulk exporters of forest products will be the Nordic countries. Afforestation programmes are likely to be reduced in the UK, and forestry projects in the Mediterranean will make little impact on Europe's stocks as a whole. The forests of Europe were calculated as being 5 per cent greater by 2000 in *The Global 2000 Report*.

An important contribution in the future must be greater efforts to re-afforest parts of the Mediterranean areas, where soil loss is still a major problem. Here, there are still difficulties with ubiquitous goats and sheep, and forestry projects may have to go hand in hand with changes in agricultural policy and increased terracing to preserve hillsides suffering erosion.

One additional factor of importance may well be the rapid increase of evidence about air pollution effects. Recent figures from Germany suggest that up to 50 per cent of the forests are showing severe signs of damage and large areas are already dead. It is very difficult at this stage to tell exactly what the outcome of the present decline is likely to be, but there has already been some wholesale forest destruction. Long, hot springs and summers, such as those experienced in the last few years, are combining with pollution to make the situation even worse than before.

It is also important to note the differences in management strategy. British visitors in Germany, or further north in Scandinavia, often find it hard to believe that the forests are not completely natural, because they are made up of many different species of different ages. Trees are felled selectively, instead of existing in rigid plantations that are clear-felled when they reach the right size. A major part of the forest debate in Europe over the next few decades is going to be about exactly how 'efficient' plantation forestry is in practice, and whether the many unpleasant side effects of it are justified.

The USSR

The Soviet Union has by far the largest timber resources of any one country in the world, with 785 million hectares of forest and an annual growth of a third as much again as the USA and Canada combined. Perhaps because of this great abundance, little care appears to have been taken with re-planting trees, and forests in the European part of the USSR are being exploited far too quickly. There is already an unsatisfied demand for timber products and a shortfall seems inevitable in the future. This is because 85 per cent of Soviet forests are in the far north, which has a very low population density and poor communications. The Siberian forests are so far removed from most Russian people that their exploitation will prove extremely expensive.

It is difficult to get accurate figures about forest loss, provision of reserves, and national parks or afforestation programmes within Russia. Sources within the Soviet Union claim that enormous amounts of timber are wasted as a result of clumsy state bureaucracy. Trees are felled and left to rot because there are no railways to take them away. Large areas are dying because of air pollution and great damage through forest fires. The anonymous author of *The Destruction of Nature in the Soviet Union* claims that between 200,000 and 300,000 square kilometres of forest and taiga are burnt every year, because of negligent and inefficient operations, and that the amount of forest burnt is continuing to increase.

A Tass report claims that re-afforestation is taking place at two million hectares per year, and implies that much of this is designated as parkland. There is no indication of how much of this programme is in plantation form. The very size of the country is an obvious advantage; regeneration in many places is likely to occur simply because population levels are so low that land management is difficult.

North America

Very large areas of forest in the USA and Canada have already been destroyed to make way for agriculture and urban development. Nonetheless, the continent still has the largest temperate forest reserves after the USSR. Both countries have over 200 million hectares of exploitable forest, although the US forest has generally been managed more intensively than that in Canada. It is also more productive, both because of its warmer climate and because of intensive management.

Canada has been guilty of very poor management in the past, and huge areas have been felled indiscriminately; in well-populated places like Vancouver Island, forest outside national parks is rapidly being logged out. In particular, the western red cedar is being destroyed throughout the island. These trees are over 1,000 years old, yet there are plans to clear-fell virtually all the old growth coastal forest. There are signs that the Canadian government now recognizes this and is planning greater management efforts; a recent official publication stated unequivocally that,

> vast areas of potentially productive forest are now inadequately stocked with trees, local shortages of wood have developed in every province and the problem of long-term wood supply at a reasonable cost is the most important issue facing the sector.

In the event, it is unlikely that any major changes will take place in the short term, despite all the rhetoric. Although some areas have been excessively logged, even a modest increase in production felling is less than growth, so that there is going to be comparatively little pressure for intensification. And large areas of Canadian forest are too remote to be exploited.

The situation in the USA seems to be far more volatile. There has been a marked polarization of views about conservation, with the big-business perspective encapsulated in Ronald Reagan's infamous quotation, 'once you've seen one giant redwood you've seen them all'! His appointment of James Watt as Secretary of the Interior was regarded by most people as a signal for uncontrolled exploitation. Watt is a fundamentalist Christian who believes that God created the earth for humans to exploit rather than to conserve, and who started plans to open up national parks to mining concessions. A storm of protest and a national 'Dump Watt' campaign eventually had him removed from office, but his period in the job saw the resignation or enforced removal of many people sympathetic to conservation from the state bodies administering forest interests. It will take a number of years to see what long-term effects current right-wing policies in the USA will have on forests.

Although economic developments will ultimately control forest policy, it does seem likely that there will be a move away from natural regeneration to a more intensive system of forestry, with a resultant decrease in ecological value, and perhaps also in long-term stability. Predictions are difficult to make at the moment and probably depend on political developments in the next few decades. Provisions for national parks are good, but the survival of some important natural forest areas, such as the rainforest in the north-west which is threatened by logging, and the Adirondack National Park trees in New York State threatened by air pollution, remains precarious.

Of particular note is the semi-tropical Everglades region of Florida. Internationally recognized as an area of outstanding ecological importance, the Everglades is protected, but its water table is not, and the numerous developments within Florida are lowering this drastically and altering the entire ecology of the region. A change in attitudes towards conservation is still quite a long way off in the USA as a whole, although 'national monuments' are often given disproportionate importance when patriotic pride is at stake.

Australia and New Zealand

Australia has about 38 million hectares of forest and New Zealand 6.2 million hectares. New Zealand is far more intensively managed and many of the forests consist of wholly alien species and single species plantations. As a result it has become more unstable and prone to sudden losses through diseases and pests. The last decade has seen an interest in natural forest and an entrenched opposition to exploitation of remaining forest areas. Plans to clear-fell one million acres of beech forest were largely abandoned as a result of public opposition. Latterly the Native Forests Action Council developed as a community response against Japanese companies involved in clear-felling southern rainforests.

The same is even more true for Australia, with two very successful campaigns to save rainforests from logging and flooding for a reservoir. Campaigns in Tasmania and Queensland have gained worldwide publicity, with massive peaceful actions against logging. The Labour Party's support for forest protection is believed to be one of the important factors in their success in the 1983 general elections (and, unlike the uranium mining issue, they appear to be keeping their word on rainforest conservation), and public support for forests is still very high. Elsewhere, of course, there are extensive deserts; deforestation and desertification are already problems in some of the more arid parts of the continent.

Many of the other Pacific islands are far more at risk. Some of the Solomon Islands have been effectively logged out, for mining and for timber extraction; the British company Unilever is closely involved in some of the deforestation. Papua New Guinea, one of the least explored areas, will have virtually all its lowland rainforest sold by the end of the 1980s, over half to Japanese companies, but also to Australia and New Zealand. While there is still extensive mountainous forest which remains uneconomic to fell, many of the richest habitats will be destroyed along with the lowland forests.

Tropical forest

Tropical forests make up about half the world's closed forests and have a growing stock of considerably more than 50 per cent, because of their faster growing rate, which is three to five times that found in temperate regions.

About 1.1 billion hectares of forest still exists in less developed countries, and this is currently being felled at a rate of about 20 million hectares per year as far as we can judge. Simple extrapolation of the currently increasing rates of loss suggests that all the accessible forests could disappear within the next 20 years if trends continue. In practice, populations and forests are not evenly distributed; large areas are still likely to exist in parts of Africa and Latin America into the next century, although their eventual survival is still very uncertain. A full country-by-country inventory is outside the scope of this book. A recent report from the Food and Agriculture Organization lists 76 countries containing tropical forests. A general overview of events in the main geographical areas is given below, along with extra details of a few chosen regions.

Latin America

Closed forests cover about 725 million hectares, over a third of the total land area. Open woodlands account for another 400 million hectares. Three quarters of Latin America's tropical moist forest is in the Amazon basin. Over the last decade, the attention of people throughout the world has been focused on the changes taking place within the Amazon, which is the last really large 'wild area' left, and which spreads over eight different countries. Until relatively recently it was widely believed to be unexploitable. Now we know better; roughly a third of the Amazon forest has already disappeared, much during the last 30 years, and currently deforestation is running at about four per cent per year.

The bulk of forest loss has been due to the spreading of settled agriculture, deliberately encouraged by Brazil, which is the largest Amazonian country and owns 60 per cent of the whole area. Multinationals have also played an important part by buying up huge areas for ranching, along with a few very rich private investors, like Daniel Ludwig, who has effectively destroyed an area the size of Connecticut in a completely unsuccessful attempt to produce plantation timber.

There is some evidence that the rate of exploitation is slowing; both ranches and farms have failed so completely that they are now discouraged. However, people with nowhere else to go are still entering forest regions in large numbers and the various governments are incapable of stopping this (at least without large-scale land reform policies that they are reluctant to implement).

The rates of deforestation vary from one country to another, as does the provision of forest reserves. For example, Costa Rica has provided good forest reserves but rates of deforestation in the rest of the country are enormous, while further south some of the Amazon countries have few reserves but still large unused areas. Although impressive areas have been put aside for national parks (Brazil alone claims to have an area the size of the UK), these are often left open to poaching and timber concessions and their real status can be very different from the declarations made by politicians.

In Central America forest is likely to be cleared entirely from arable land by the turn of the century, although there will be a few reserves and forests in mountainous regions. Large-scale ranching has played an important part in Central American countries and many cleared areas have already reverted to scrub. Proposed schemes to make savannah land arable will, if they work, take some of the pressure off remaining rainforest areas but will mean further clearance of open woodlands in Argentina, Paraguay, Bolivia and Brazil.

Planted forests cover about two million hectares in Brazil and 1.7 million hectares in the rest of the continent. Planting in Brazil is apparently increasing and a wider variety of species are being used instead of the ubiquitous eucalyptus. In the drought-ridden north-east, over a billion trees are to be planted in a massive afforestation programme planned to re-establish the forest and change the local climate. Ten per cent of any plantation area is supposed to be set aside for a recreation of the original forest. With good management, South America could easily rely on plantation timber without clearing any more primary forest, but this is very unlikely to happen.

The ecological effects of mass clearance of tropical forests in Latin America will be disastrous unless deforestation is curbed soon. Two thirds of the world's freshwater flows through the vast Amazon hydrological system and half the rain which falls there is evaporated, largely through the vegetation. The soil is poor and only a small area will ever be suitable for agriculture. Biologically speaking, the Amazon is both the richest and the least known region on Earth. Because of its evolutionary history, a number of lowland areas have exceptionally high concentrations of species, so-called 'biogeographical islands', and the preservation of these is now considered to be of primary importance. Continued forest destruction will certainly destroy species of plants and animals, ruin the soil over huge areas, and alter the climate.

It is difficult to judge exactly what will happen to the South American forests, and simple extrapolations from current events are likely to prove inadequate. Almost all the deforestation factors described in this book apply to this continent; continuing political instability there is likely to make the situation even worse. On the other hand, the countries of South America have learned some hard lessons over the past ten years and there are signs that a more mature forest policy may emerge in the future.

Africa

With only 6 per cent of its area covered with closed forest and perhaps three times that amount of open forest, Africa is the Third World continent with the least forest, largely as a result of felling in the near and distant past. Much of what's left is in the tropical forest regions of west and central Africa. Estimates of destruction are notoriously difficult, but it is thought that at least two million hectares are lost every year, and the real figures could easily be twice this or more. *The Global 2000 Report* stated bluntly:

Assuming no changes in current patterns of land use or in priorities for development project funding, substantial areas that are covered now with tropical moist forest will become barren wastelands with soils that have no potential for production, and many areas that are dry open woodlands will become deserts.

Closed forest is expected to be reduced to at most 146 million hectares by the year 2000 and may easily have been reduced to 130 million hectares if felling increases.

The main factor in clearance is agriculture, and this could become far worse if the tse-tse fly eradication programmes are developed, because these rely on the clearance of timber as one of the prerequisites of disease control. There is no proper survey of the rate of savannah degradation, although the recent interest in desertification has focused some attention on this problem. Deserts threaten some 7 million square kilometres in Africa south of the Sahara, with more land at risk in the northern African countries as well. Permanent agriculture is likely to take large areas of forest for crops like coffee, cocoa and palm oil in the next few decades.

Most of the felled timber (85 per cent) is used as fuel; this has increased since the oil crisis raised the price of kerosene and is likely to continue due to expanding populations. Already most African countries are importers of timber and a shortfall seems inevitable in the near future.

Some of the largest areas, like Zaire, the Congo and Gabon, are still relatively untouched, while forests in countries like the Ivory Coast, which is currently the largest timber exporter in Africa, will only survive a few more years. Virtually all the forests in Ghana and Nigeria have now been marked for exploitation, and increased access to forests in the future is going to put more areas at risk. Even north of the Sahara, where most of the trees have already gone, there is a continuing loss of trees and replacement by exotic species in coastal areas.

Although African forests seem less threatened than in the other two continents at present, they are also smaller and the situation is changing fast. The fact that Zaire, with a notoriously unstable government, has control over the largest area, is itself a factor of considerable concern.

Asia

In terms of immediate threats to large forest areas, Asia is probably the most badly affected continent. While there are still some extensive forest tracts remaining, huge areas of lowland forest have already been logged out, or will be within the decade. Very large areas are likely to be replaced with plantations, others are left to shifting agriculture. Large-scale soil erosion problems have already resulted.

The richest tropical rainforest areas are located in Indonesia, Malaya and the Philippines. India, China, Burma, Thailand, Korea and other Asian countries tend to have more extensive areas of mixed and semi-dry deciduous forest. Areas for the less developed Asian countries are about 135 million hectares of rainforest, 55 million hectares of tropical moist deciduous forest, 95 million hectares of other deciduous forests and 50 million hectares of coniferous forest.

There are several crisis points in Asia. The forests of the Philippines, Malaya and Indonesia were probably the richest habitats in any area of comparable size in the world, having evolved undisturbed for 60 million years. Human activity in the last few decades has effectively logged out (i.e. extracted some timber and destroyed more in the process) vast areas of the Philippines and Malaysia. Pretty well all the lowland forests will have been logged in these countries within 15 years; Indonesia will follow the same course not long afterwards. These countries, along with China, Mongolia, Burma and Thailand, have been exporting large quantities of timber, mainly to Japan and Western Europe.

Once logged, large areas will be left to regenerate, although it is by no means certain how well forests will recover in practice. The delicate fabric of lowland primary forest has already disappeared over most of the region. Other parts will be replaced by plantations and more by agriculture. In the Philippines an estimated 200,000 hectares of forest disappears every year because of shifting agriculture, while 250,000 hectares go in Thailand, despite the imposition of the death penalty for illegal felling. Indonesia already has about 30 million hectares of degraded grassland as a result of deforestation and a further 2 million hectares are being farmed in an unsustainable manner.

In Asia as a whole, most of the forest is lost to shifting agriculture and most of the wood is used for fuel. Further west, in India, Tibet and Pakistan, deforestation in the uplands is also causing grave problems with flooding and fuel shortages. Forest cover on the watersheds in the Indian Himalayas is down to a quarter of what it should be, which is far too little to hold back the monsoon rains and release them gradually as in the past. In autumn 1978, a flash flood inundated 66,000 villages, killing 2,000 people and 40,000 cattle. The value of lost crops in two states, West Bengal and Uttar Pradesh, was put at $750 million and annual expenditure for flood damage has averaged $250 million for the last 30 years. Soil is degrading faster in these areas and 1.7 million hectares of India and Pakistan are threatened with desertification.

There have been some spectacular replanting efforts in Asian countries, notably South Korea and China, although in the latter case planting for watershed production has met with very widespread seedling failure. FAO estimates that plantations cover 30 million hectares. Plantations are expected to cover 90 million hectares of Asia by 1990.

By 2000, natural closed forests may well be reduced to 100 million hectares, mainly in protected or inaccessible areas. Most of the rest will be degraded through timber extraction, while other areas will become plantations. Deforestation, desertification, siltation and flooding will have severe effects on food production and energy supply. *The Global 2000 Report* concludes soberly that 'it seems likely that some of the region's present potential to support human population will have been irretrievably lost.' Much of the wildlife and forest resources will also have been irretrievably lost in the process.

Tropical forest loss: an overview
The sections above are necessarily brief and superficial. I hope that they help provide a general picture of what is happening in the three major Third World areas. It must be stressed that all these figures are still approximate. Local estimates can be wildly innaccurate and there is also a tendency to give false figures, for a whole variety of reasons. The use of Landsat photographs is

helping to untangle fact from fallacy, but while the overall accuracy of estimates may be improving, there are still many regions of uncertainty. In particular, it is almost impossible to assess the damage that logging has done, especially in areas where trees are still standing, and so show up as forested on Landsat prints.

As a kind of appendix to the tropical section, the findings from Norman Myers's survey of tropical forests are reproduced here in tabular form:

Areas undergoing rapid felling (i.e. much gone by 1990)	Areas undergoing intermediate felling (i.e. much gone by 2000)	Areas apparently still stable
Philippines *	Papua New Guinea *	Brazil (parts of)
Malaysia *	Burma	West Africa (parts of)
Indonesia *	Colombia coast	Zaire basin
Melanesia *	Ecuador (Amazon)	
Australia **	Peru (Amazon)	
Thailand	Brazil (parts of)	
Vietnam	Cameroon (parts of)	
Bangladesh		
India		
Sri Lanka		
Central America		
Colombia		
Ecuador coast		
Brazil (parts of)		
West Africa (parts of)		
Madagascar		

  * especially lowland forests
** now more action to protect rainforest

Source: Norman Myers, 'The present and future prospects of tropical rainforests', Washington D.C.: Environmental Conservation 1980.

It would be unfair to say that the picture is totally gloomy; several countries now recognize the problems and are taking active steps to try to reverse the process. But with mounting debts, civil unrest, insufficient forest funding and widespread political corruption, making any headway in the timescale required is not going to be easy, and we have to accept that many more areas are going to be deforested and converted to unusable land in the future.

8. Behind the headlines: some real causes of forest abuse

Too many of the accounts of deforestation stop here: blaming forest abuse on seemingly irreversible factors or on individual greed and stupidity. We tend to classify the problems into geographical areas, with those in the 'Third World' being neatly separated from those in the north. When wider problems are discussed, the analysis tends to be simplistic, so that problems are thought to be due to 'overpopulation', or to the excesses of individual companies. But a more general analysis can be made.

In the section below, an attempt is made to look at some of the broader issues affecting forests. The question of population increase is tackled first, as this is one of the most discussed, and misrepresented, of all 'development problems'. This leads into a more general discussion about access to land and the emerging role of capitalism in land use. The part played by transnationals is also covered and the section finishes with a brief look at the part that cultural roles play in people's – especially men's – attitudes to forest exploitation.

Population

The importance of population growth as a factor in the problems facing Third World countries is one of the most contentious issues in the development debate. 'Overpopulation' has been used as a universal explanation for poverty, famine and ecological damage by reactionary theoreticians in both the north and south, and has proved a convenient way of shifting the blame for the problems of poor countries on to the people who live there. Indeed Thomas Malthus, one of the first people to draw attention to the dangers of population growth, specifically outlined the risks of uncontrolled expansion of the lower classes! This general feeling, whilst usually put rather more tactfully, still pervades most literature about population, and many well-meaning people are convinced that the only hope for the Third World is for the more civilized nations to teach (or enforce) population control measures in areas with high birthrates and little space.

Not surprisingly, the left or liberal development lobby has reacted against this 'white man's burden' approach, looking behind the population issue at the real reasons for inequalities and poverty. Writers like Susan George, Frances Moore Lappe and Joseph Collins have argued persuasively that hunger has very little to do with population and a lot to do with the buying power of the north, so that in a poor country people can be starving right next to a plantation growing food or cash crops for export to a rich country. Other writers have argued that environmental destruction has more to do with the greed of transnational companies than with the breeding habits of aboriginal people; the population problem has tended to be written off as a complete red herring.

The truth lies somewhere between these approaches. The myth that all (or almost all) development problems are due to over-population is at best naive and is often simply a cynical way of unloading responsibility for the problems on to people without resources to argue back effectively. But ignoring population growth altogether is also dangerous, particularly on a more local scale where a sudden rise in population can have disastrous effects on food reserves and ecology. Over the past two or three decades, forest areas in the tropical countries have seen a very rapid increase in their human population due both to a rise in birth rate and to an influx of people from elsewhere. If we are to understand the real impact of human population on trees it is important to have some idea about the dynamics of this increase.

People in forests

There has certainly been an enormous expansion of people in tropical forest areas over the past 20 years. Foresters estimate that something like a quarter of all tropical moist forests are now inhabited by 'permanent' farmers, which as we have already seen, are actually likely to be fairly temporary in practice. In the tropical deciduous forests of Asia, areas that used to provide firewood at a sustainable population density of one person per hectare, now have up to 15 people trying to collect firewood from each hectare. Villages in India that were surrounded by thick jungle only 20 years ago now stand in scrubland area with not a tree in sight.

On a micro-scale, population pressures are causing enormous problems for forests, and in turn, for the people trying to eke out a living in forest areas. However, it is still not clear how and why people have come to be there.

Population growth

Populations are growing very fast in many developing countries, and this is bound to cause stresses and problems in both urban and rural areas. India's population is doubling every 17 years at the current rate of expansion. Similar growth patterns are occurring in several other Asian countries and, to a lesser extent, in parts of Africa and Latin America. Larger families need more food and more fuel, so to some extent deforestation is bound to be linked to growing numbers.

Except for the possibility of nuclear devastation or some other huge calamity, there is little doubt that the world's population is going to continue to expand very rapidly for the next few decades at least. *The Global 2000 Report* quotes a medium estimate of a 55 per cent increase from 1975 to 2000, i.e. from 4.1 billion to 6.35 billion. Roughly 90 per cent of this is expected in the less developed Third World countries, which would then contain some 80 per cent of the world's population. The official UN medium-term projections show world population levelling off at about 10.5 billion 100 years from now, that is, considerably more than twice the current numbers.

There is absolutely no doubt that the pressure of extra numbers of people will add greatly to the suffering of those already living in abject poverty. This will be accompanied by further environmental degradation and hence further human misery. While it is theoretically possible to feed and clothe this number of people perfectly adequately, the political climate, the distribution of people, and personal and national greed give little hope that this will occur in the foreseeable future. China still leads the world in overall growth of population, despite some draconian efforts to contain it, but other countries have now overtaken it in percentage increase, including Mexico, Brazil, Pakistan, Bangladesh and Nigeria.

The two principal causes of deforestation are, as we have seen, clearance for agriculture and collection for fuelwood. Ninety per cent of people in the poorest countries (some two billion people in all) depend almost entirely on wood for energy, and increasing numbers of people, particularly in areas suffering deforestation in Asia and Africa, are having a direct effect on forest clearance through their energy requirements.

Whilst it is not really within the scope of this book to look at the reasons behind population growth, it is worth pointing out briefly that an expanding population is not simply a sign of 'ignorance', to be countered by severe population control measures. Indeed, apart from people with strong ideological incentives, like the Chinese, most imposed population control measures have not worked well in practice. The desire to have large families is tied up with many factors including status, tradition and a very pragmatic desire to provide insurance against one's old age in countries without provision for pensions or social services. Legislation seldom works and publicizing contraception does little good in societies where women are unlikely to be allowed to take their own decisions about bearing children. Population stabilization programmes have only succeeded in those places where the introduction of birth control has gone hand in hand with the emancipation and education of women.

Population movement

However, it is a dangerous over-simplification to assume that just because population growth is an important factor in some areas that it is the complete story. Notwithstanding the fact that growing populations are causing serious problems in some areas, there is still a great deal of under-utilized land in many Third World countries, well away from the forests. The sheer expansion of numbers is often not the real cause of the problem. Rather it is often that people are being forced into forests from other more desirable land areas.

The past two decades have seen a quite unprecedented increase in migratory and landless peoples. They have been forced to move because of religious and racial persecution, because of unacceptable political systems, because of starvation, drought, crop failure and lack of food, because they need work and, increasingly, because they are forced off traditional land by lack of money or outright threats of violence.

There are, very roughly, about 16 million refugees currently on the move or in temporary accommodation throughout the world, many in Third World countries. Although a few mass migrations receive the bulk of media attention, like the Vietnamese boat people, the refugees from Pol Pot in Kampuchea, and the Palestinians, there are in fact constant movements of peoples between many countries.

However, apart from a few propaganda stunts, like a capitalist country welcoming fugitives from communism, and vice versa, most refugees are not welcome. They are often reliant on funds from Third World governments which are already overstretched. Refugees' motives are often seen as suspect by the receiving government, especially if they 'escape' to a richer country. And harbouring refugees can pull a neutral country into a conflict with which it has no desire to become involved, as troops cross borders to attack suspected guerrilla strongholds in settlement camps.

In a few cases, refugees are only in other countries for a fairly short time and, apart from the economic costs, their presence will do little permanent damage. The largest recent case of repatriation involved the ten million Bengalis who had fled to India during the Indo-Pakistani war and were later returned to the new nation of Bangladesh in 1972. However, in many other instances refugees may remain in limbo for years or decades. If international aid agencies cannot provide adequate funds, these people will eventually be forced to integrate with the host country, or move again. Many of these luckless people end up in the least usable land areas. These often include the forests.

Land ownership

A much more important cause of landlessness in Africa, Asia and Latin America is the continuing concentration of land ownership in the hands of a few people. Patterns of land ownership have always shaped patterns of human relations and it is always the landless or near-landless that are at the bottom of any social scale.

Landlessness is now endemic to many areas of the Third World and, in countries that still rely predominantly on agriculture as a livelihood, ownership of land is the real key to political power. Landless labourers, sharecroppers and marginal farmers now make up the bulk of the population of many less developed countries. In Asia, the proportion of families that are almost or completely without ownership of land ranges from a minimum of 53 per cent in India to 85 per cent in Java. These figures are paralleled in Latin America by 55 per cent in Costa Rica and 85 per cent in Bolivia and Guatemala. Just seven per cent of the population owns 93 per cent of agricultural land in the Latin American continent, according to FAO. In all, some 600 million people live in rural households without secure access to land. It is no coincidence that this figure approaches the 800 million who, as the World Bank estimates, live in 'absolute poverty . . . at the margin of existence'.

Whilst there have always been gross inequalities in different parts of the world, today several factors are combining to make the situation worse than before. Until recently, even very rich landowners needed to keep large numbers of people on their land in order to work it. Whilst these peasants often had miserable and degrading lives, they were at least fed sufficiently to keep them alive and healthy enough to work, and were also frequently given somewhere to live and some land on which to grow their own food. Today, machinery has taken away many of the incentives for keeping people on the land. A farmer or forester can produce the same or better crop through the use of capital-intensive machinery and expensive agro-chemicals, with far less labour.

This has not only displaced people who would have traditionally found a living on the lands of the rich, but it has also displaced the smallholder existing alongside the big farmers. Twenty years ago, the 'Green Revolution' was launched as the way of saving the world from starvation. New crop varieties, bred in the laboratories of the north, were exported to the Third World in enormous numbers. These varieties did indeed give higher yields, and for several years agronomists talked proudly of increased food supply and an 'end to hunger'.

Unfortunately, the plant breeders ignored the politics of the development. The new crop strains might give higher yields, but they also needed higher inputs of fertilizer, pesticide and herbicide, supplied by the chemical companies from the same countries which had pushed the Green Revolution so enthusiastically. The results for the small farmer were disastrous. Having insufficient money to buy expensive agrochemicals, smallholders had to continue growing their traditional crops. Meanwhile the larger landowners could invest in the necessary pest control and produced far more food than before, forcing market prices right down. Facing bankruptcy, the small farmers were forced to sell to their richer neighbours and were soon unwanted even as labourers. Meanwhile, the rich survivors frequently found it was more profitable to export cash crops, like tobacco and coffee, to the north than to grow food for their own people at home. Far from solving the food crisis, the Green Revolution has almost certainly intensified it, in some cases.

When this happens, Third World governments have tended to try and find somewhere else to put their landless people, rather than to face up to the necessity of land reform. In several forest countries, this has led to the settlement of people in jungle areas, along ribbon developments bordering on roads and around mining areas. Government-backed 'resettlement' schemes have been forced upon desperate people with nowhere else to go. Faced with the forest or starvation, hundreds of thousands have tried to till the unworkable jungle soils into some semblance of agriculture. As we have seen earlier, their efforts have led to disaster.

There is now considerable agreement among governments that resettlement schemes in tropical forest areas simply do not work, and several countries have tried to restrict access to forest areas. Unfortunately, very few of them have faced up to the fact that they have to provide somewhere else for people to go. Those that have tried to redistribute the land have usually lacked the political power to take good land from an entrenched and powerful elite.

For a growing number of countries, the increasing polarization of land ownership has provided the final trigger for socialist revolution, in South East Asia, North Africa and Central America. Throughout the Third World, tens of thousands of peasants have lost their lives fighting for land during the present century. The need for land reform has been well publicized by the United Nations, the major aid agencies and all the principal charities as well as by the left. Most Asian and Latin American countries now have some laws about land reform and officials swear solemnly that procedures are going ahead. And so they may be, but unfortunately nowhere near fast enough. National laws are delayed, watered down, circumvented or, if all else fails, simply ignored by the rich and influential. Land ceilings (i.e. a ruling on the maximum amount of land under control of one person) are avoided by dividing land up among family members. Tenancy reforms are often avoided by the simple expedient of evicting tenants. International pressure for land reform from the west is often influenced by the fact that land reform is promoted by socialist groups in the Third World; these are, of course, viewed with misgivings by many western governments.

There are far more practical reasons for the west's ambivalence towards land reform than simply paranoia about communism. Land reforms cut into the rich nations' interests, both by expelling transnational corporations or by reducing the availability of cash crops in favour of subsistence foods. As Erik Eckholm has written, 'experience indicates that the lofty goal of land reform is readily sacrificed on the altar of global geopolitics.'

There have been reasonably successful land reforms, but they have rarely been carried out in a gradual or peaceful manner. The sweeping programmes of China, Cuba, Mexico, the Soviet Union and Vietnam, for example, have been the direct result of a successful revolution. In another group, which includes Japan, South Korea and Taiwan, they have been implemented by right-wing governments in the aftermath of a political war, as a way of reducing unrest.

Gradual reforms have sometimes been partially successful, for example in India, Iran and Venezuela, but dozens of countries have formulated laws and completely ignored them thereafter. A farsighted paper published by the Agency for International Development in 1970 stated bluntly:

Only in those cases where political leadership or political opposition makes a definite decision to build an articulate peasant political base is there likely to be significant new land distribution . . . the tragedy of looking to peasant mobilization and organization in most countries is that . . . the forces of reaction and suppression are great and often severe . . . Perhaps in many present less developed countries such peasant mobilization is inherently and unavoidably revolutionary.

Capitalism

The popular western view of poverty-stricken Third World countries is one of many small farmers trying to eke out an existence for themselves and their families in drought-afflicted or overcrowded land. As we have seen, the small farmers' position is in reality constantly being undermined, and a more accurate picture is of a mass of landless peasants trying to survive at the edges of sparsely populated agricultural land owned by a tiny elite. Many landless peasants are then forced to retreat into deserts, scrub and jungle.

Much of the agricultural land in the less developed countries is not used to grow food for home consumption, even where the majority of the population is malnourished. Most good farmland in the south is set aside for producing cash crops to sell to the north. Vast areas are used for the production of best-selling commodities which have little food value, like coffee, cocoa, tobacco and sugar. In the El Chapane region of Bolivia, 75 per cent of the land is used to grow coca to manufacture cocaine, adding to an 'invisible export' of $1-2 billion a year, mostly going to a few enormously rich gangster barons. In Costa Rica, where vast forest areas were cleared to grow beef, production increased by 92 per cent during the 1960s but home consumption fell by 26 per cent in the same period, the rest being sold abroad, mainly to the USA.

Cash crops are by no means new to the Third World of course; sugar cane has been grown in Brazil for over 400 years for example, and export crops have been a major factor in both the economy and the amount of political unrest ever since the European expansion. However, in the last few decades the demands of the rich countries have grown steadily larger and, perhaps more importantly, sales of cash crops are increasing in poor countries as well.

It is worth emphasizing this last point. Far from just switching people from one brand of food to another, the companies operating in the Third World are creating a whole new consumer class. The promotion of consumer products persuades peasants and subsistence farmers to switch from food to cash crops, to raise money for items like soft drinks and cigarettes. While it can be argued that these people have every right to 'luxury goods', in the short term a switch to more cash crops can lead directly to malnourishment of people and degradation of land. This development of cash cropping has had a direct effect on the survival of forests and open woodland, and particularly on the spread of deserts in many parts of the world. It causes damage to the land when a traditional crop, which is well-adapted to a particular habitat, is replaced with an imported cash crop which is not. In the African Sahel, the expansion of groundnut production at the expense of drought-resistant crops like millet has been a contributary factor in desertification. Groundnuts are susceptible to drought, so that replacing traditional crops with them can be disastrous. Accordingly, farmers have planted them on fallow land, reducing grazing space for goats and other livestock. When grass disappears the animals are forced to eat the leaves and twigs from trees which themselves die, taking away shade for other plants. The desert spreads.

It is obvious that at an individual, micro-scale level, capitalism is having a detrimental effect on the environment. Taking a broader view, the same is true on a national scale. Timber now ranks among the top five exports from the Third World, but many countries are exploiting timber in an unsustainable manner. It is wholly unreasonable to expect debt-ridden countries to abandon their timber trade when it is one of the few commodities they have to sell. Nevertheless, this commodity itself will disappear before very long unless changes in the attitudes to forests take place. The economic pressures exerted by the north are forcing Third World countries into an unsustainable 'production' of timber, and it seems as if it will be impossible for them to avoid a crash in the future.

Transnationals

Such firms, some 500 to 700 in total, mainly American, now account for the bulk of direct investment, production, trade, finance and technology of the non-socialist world. About 200–300 multinational companies (MNCs) account for over 70 per cent of US foreign investment, about 165 for 80 per cent of that of the United Kingdom and 82 for over 70 per cent of that of West Germany. Much of this is by the largest firms – the top 50 – whose relative size has tended to increase over the last decade. In terms of value added, production by all MNCs now exceeds one fifth of the total of the gross national products (GNP) of the non-socialist world. The production of each of the leading MNCs is greater than the GNP of over 80 countries. (Sanjaya Lali, 'Multinationals and Development', *National Westminster Bank Quarterly Review* February, 1975.)

The role of transnational corporations in Third World development is one of the most contentious issues in the development debate, with a sharp divergence of opinion about their advantages and problems for poor countries. From our perspective of forests, two side effects of transnationals are of overriding importance. First, many are interested in the control of raw materials; they play a major role in exploiting (and over-exploiting) forests, both by taking timber out and by clearing land for ranching. Second, their global mobility means that they can move in fast to exploit areas as they become available, and then get out again just as quickly when conditions change. These factors have led many transnationals to become involved in forest exploitation in an irresponsible and unsustainable fashion. At the same time, their actions can be completely logical from their own perspective.

Take a hypothetical company, 'Treefell Inc', which has bought a large timber concession in a densely forested and under-explored tropical country. It is unlikely that there will be roads into the forest region, so Treefell will have to build them, usually using local workers administered ultimately by white managers trained in the north who have no long-term commitment to the country. In the humid, disease-ridden jungle the costs of road construction are high, even though they are not generally planned for long-term use and labour is very cheap. In these circumstances, Treefell will not react enthusiastically to suggestions of re-routing the road to avoid a newly discovered aboriginal group. An appeal directly to the government (and some judiciously applied pressure and bribes) helps them to get round this, although periodic attacks by tribesmen necessitate the extra expense of armed guards for the construction workers. To make matters worse, a sawmill will probably have to be built and an enormous amount of extremely expensive equipment imported.

Once established, the concession can either be exploited gradually and carefully, or fast and recklessly. The final profits of the former method will be greater in terms of total amount of timber extracted, and the forest will also be in a far better state to regenerate. However, two important factors make Treefell cautious of the 'softly softly' approach. First, the high investment costs and interest rates mean that it is important to get a return as soon as possible. A smaller amount of timber which can be sold more quickly may well make better economic sense when the money can then be invested somewhere else, quite possibly in another country and in a business totally unrelated to timber. And secondly, the notorious political instability of much of the Third World means that many companies plan only a few years in advance, and are constantly afraid that a swing to the left in countries where they operate will result in greater controls over their activities, or even expulsion.

It is, therefore, in Treefell's interests to get as much timber out as possible, as quickly as possible. This means concentrating on the most valuable tree species, taking as many as possible from any area and not bothering too much about those left behind. The result will be a badly degraded forest, which will take many years to recover and could lack the most useful species in the future. Some areas will never regenerate and will end up as scrub. But by then Treefell will no longer be working in the country, or at least not in the timber trade.

For the ecologist, it is obvious that Treefell is acting with supreme irresponsibility. But from the point of view of the company and its shareholders, it is acting in the very best way that it knows, at least in the short term. By taking the fast approach it is minimizing the risk to its investment, providing capital profits which can then be invested elsewhere and keeping ahead of its competitors in the timber market.

The companies
Over the last few decades, firms from Europe, North America and Japan have all moved into tropical rainforest areas, practising

logging, plantation forestry, ranching and mining. About 50 large US firms are now involved in tropical countries, with a slightly smaller number of Japanese transnationals and a scattering of firms based in Britain, The Netherlands, France and other European countries. While many of these are directly concerned with the timber industry, others only have a second hand interest in terms of the food or minerals which can be produced from the land. A third group includes firms which have no previous interest in forestry but are simply expanding their operations into new areas.

The loggers
The effect of logging firms has been sketched out earlier. Whilst theoretically loggers can leave an area suitable for natural regeneration, in practice their scramble for quick profits has frequently meant that the degradation is long term and often permanent.

Take the American firm Weyerhauser, for example. Established in 1900, Weyerhauser is one of the largest US firms, owning about two million hectares of forest in the USA and extensive tracts in Canada. It employs some 50,000 people and had annual earnings of $230 million in 1981. Weyerhauser has been involved in concessions in three major tropical forest countries – the Philippines, Malaya and Indonesia – although it has since withdrawn from some of these. Researchers have accused the company of essentially buying its way into tropical countries; in Indonesia the company owned a 65 per cent share in one of the concessions, and the Indonesian army owned the remaining 35 per cent, but the Indonesian partners reportedly supplied nothing more than the licence. Weyerhauser put up all the capital and expertise.

The concessions have not been run on a sustainable basis. As a company representative told *Environmental Action* magazine in March 1980: 'We don't yet know the downstream results of tropical logging. We just can't answer definitely about whether a dipterocarp forest can be successfully regenerated.' Deforestation caused severe flooding in 1977 and 1979, destroying rice fields and resulting in disaster for tens of thousands of local people. Tribal groups have suffered; traditional rights to collect food and materials from forests have been lost. Nonetheless, the logging continued, and Weyerhauser even started to look at ways of replanting the area with fast-growing exotic species. However, the Indonesian government began to demand the regeneration of slower-growing native tree species, whereupon the company sold out its share in 1981.

The US giants involved in tropical forest exploitation include Boise Cascade, the Scott Paper Company, Crown Zellerback, Dixie Wax Paper, Potlach, West Virginia Paper Corporation, and over 30 other companies operating in Latin America and South East Asia. Japanese firms are active in Asia and Australasia, including concessions in Papua New Guinea and even Australia itself. A survey of nine transnational companies in Indonesia showed that not one was following the regulations laid down regarding maximum trees taken per hectare and care in felling and extraction.

The largest British multinational involved in forestry is Unilever, a joint British and Dutch based firm, which has caused much anger from local people in Asia and the Pacific for having logged out whole areas and islands, having failed to replant cleared areas and having allegedly destroyed coconut plantations in the process. Despite a well-funded public relations exercise illustrating all the beneficial results of Unilever's Third World activities, there is growing opposition to their activities in the field.

Ranching
The extremely short-term option of stripping rainforest to create cattle ranches shows transnational companies at their most avaricious. Agribusiness barons, including Swift-Armour Meat Packing Company and King Ranch, are joined by giants like Gulf and Western, Del Monte, Goodyear and United Brands. The largest ranch yet attempted, some 840 square miles, belongs to the German car manufacturer Volkswagen. The pall of smoke created when the forest was burned to create pasture was visible by satellite picture, generating stern censure and a fine from the Brazilian government.

One of the most contentious issues is the role that American fast-food chains have had in ranching. While companies like Burger King and Jack In The Box admit to using imported beef, other chains, sensitive to the growing anger about deforestation, have sought to dissociate themselves from the ranchers, issuing public statements stressing their lack of involvement with the trade. Some of these do not stand up to very close examination. McDonald's, by far the largest fast-food chain, with branches in over 20 countries and among the top 150 corporations in the world, categorically denied that it uses beef from outside the USA. On-the-spot investigations in Central America showed that beef from Guatemala and Costa Rica was sold to McDonald's. The discrepancy is possible because the Meat Importers Council of America categorizes any meat imported into the USA as 'domestic beef', so that McDonald's can legally say that they 'obtain their meat within the United States', whereas it was actually raised in the ashes of tropical rainforest!

While some companies are now pulling out of tropical forest areas, for a whole number of economic, strategic and political reasons, others are still very much involved. Just as worrying, Third World transnationals are now growing up to fill the gap left by any departing western corporations. These are playing an increasing role in the world market. However, the bulk of transnationals are still found in North America, Europe and Japan; by far the most important role is played by the USA. For every dollar that US companies invest in Latin America, three dollars come back in profit. As Nathaniel Davies, the US ambassador in Guatemala said to the US Chamber of Commerce in 1971, 'Money isn't everything. Love is the other two per cent. I think that characterizes the United States' relationship with Latin America'.

Masculinity and forest destruction

We have seen that forest destruction is often the result of greed rather than necessity: individual greed, which results in all the good land being controlled by just a few people, and corporate greed, which encourages companies to rip out forests for short-term profits. However, it would be an oversimplification to suppose that the story begins and ends with accumulation of capital. There are other, more subconscious reasons why people want to push into new territory, such as 'taming' wild jungle and imposing their own designs on enormous tracts of land. A large part of the drive to exploit rainforest in at least some areas is as much a factor of masculine pride as of business acumen; men want to clear forests because the lumberjack is the tough guy who pulls all the women.

One of the most obvious examples of the link between forest clearance and machismo can be seen in the current ranching practices of Central America. Many 'ranchers' are actually businessmen and lawyers, who spend the week in the city and retire to their haciendas for the weekend. For them, clearing the virgin jungle puts them firmly on the frontier, as they watch their *peons* beating back the wilderness before retreating back to their air-conditioned houses for the rest of the week. They don't want to use the already cleared land, because farming is not so exciting; the real cowboys are the ones working on the edge of the wilderness.

The same kind of syndrome can be seen in miniature within the more developed countries. The invention of the chainsaw has meant that felling trees is quick and easy but, perhaps more importantly, it has brought the 'image' of the forester directly into the latter half of the twentieth century. Instead of a crude axe (which requires considerable skill, as well as effort, to use) the man felling a tree today has a complex piece of machinery. In the same way that the technical expertise used in driving a motorbike has a more 'masculine' image than the physical power needed to ride a bicycle, the chainsaw has replaced the felling axe as a status symbol and has opened the door for many more men to go out and cut down trees at the weekend. It is doubtful whether the current boom in woodstoves would have taken off in the same way if people still had to cut wood by their own efforts. (In the areas where this is still the case, it is usually the women who collect the fuelwood anyway.)

Unfortunately the same thing can happen on a far larger scale. The vast timber operations in the Amazon may have been aimed at profit, but they also have the indefinable glamour of 'man against the wilderness' as an added incentive. Conservationists tradition-ally steer clear of the troubled waters of sex roles as vigorously as they avoid politics, but the two are often very closely linked and should not be ignored if real advances in understanding the psychology of environmental destruction are to be understood.

9. Forests for the future

Norman Myers has drawn an important distinction between the two philosophies with which we exploit forests. Forests are *over-exploited* in the way they are felled, burned and otherwise destroyed, but *under-exploited* in that we have not even begun to utilize more than a fraction of their resources in terms of food, healing potential through drugs, genetic material, and other valuable products. Any long-term strategy for forests must include a switch from an unsustainable type of exploitation to one which concentrates on sustainable management and harvesting.

There are certainly not going to be any quick or easy answers to forest abuse. More land is going to be degraded, more deserts formed and more ecologically unsound plantations established in the future, whatever happens over the next few years. However, there are signs that the reckless exploitation of the forests will be challenged more vociferously than has been the case in the past, and that more balanced forest strategies are gradually emerging.

There are a number of possible fates for the remainder of the world's forests:

- Forests can be left essentially untouched, so that wildlife can survive into the future and aboriginal people can carry on their lifestyle indefinitely, or at least for as long as they wish to.

- Forests can be exploited in a sustainable manner, either by managing existing species or by replanting with introduced species, probably in plantations.

- Trees can be clear-felled and the land converted to agriculture and other uses.

In addition there is a fourth possibility, that land which has already been cleared will be re-afforested, either in plantations or as mixed, partially regenerated forests.

All of these activities will, in fact, take place. The important questions at present are how much land will be cleared, how much of this cleared land will degrade to scrub and desert before practical management plans can be adopted, and how can sustainably exploited forests be managed in practice. Well-meaning global strategies are unlikely to be of much use to many of the countries struggling with acute forest problems over the next few years. Their own political development, pressures from the industrialized countries, the expectations of their people and the changing role of timber in the world market are all important factors which are bound to change in the future. In practice, environmental policy will probably continue to be a sometimes confused mixture of ideology, practicality and vested interests for the next few decades, at least until an ecological perspective is more firmly grounded into the minds of people and governments.

Priorities for saving the world's forests

1. International aid, finance and pressure for the establishment and management of forest reserves. The crisis for forests in many countries is occurring right now, and long-term plans for the years ahead are going to be of little use if resources have already disappeared. There are currently too few forest reserves, and those that do exist are frequently little more than paper declarations. Areas are still exploited despite their 'reserve' status. It is often difficult even to find out where they are, let alone to expect them to survive intact. Additional political pressure needs to be backed up with extra money from the richer countries to help maintain them. The recent initiative of setting up independent reserves, funded from outside a country, is a positive advantage when the country owning the forest area is already desperately poor. At the same time, a far more responsible attitude must be taken towards forests in the rich countries as well, especially to those in the UK, where the population is high and surviving areas of ancient woodland are extremely limited.

2. Greater control of transnational companies is essential, although it is likely to cause some schisms within the 'conservation' movement, and anti-transnational campaigners are likely to find more help outside the traditional environmental field, such as from development activists interested in food and health. The whole issue of the power and influence of transnational companies is a complex one requiring a broad-based political response, but in the short term the following demands are vital if forests are to thrive:

* far more care in extracting timber so as not to damage trees which are left in the forest.

* provision for the replanting of trees in cleared areas or – perhaps better – time for natural regeneration to take place, with replacement of those commercial species which have been extracted.

* encouragement for governments to ban more obviously wasteful practices like large-scale burning of forests for ranching and clear-felling for plantations.

* international pressure to prevent logging companies exploiting timber concessions within national parks and nature reserves.

Pressure for these moves is unlikely to come from governments in the north (at least initially). Environmental groups and consumer organizations will have an important role to play here.

3. Greater international co-operation in forest management. The setting up of multi-state forest bodies in the Amazon and Africa is an encouraging sign that co-operation is possible. One logical development from this would be the setting up of an international cartel, like the oil-producers' OPEC, to help standardize prices; this would not be easy in practice because the north is comparatively affluent in timber resources, giving the countries which would really benefit from a cartel a poor bargaining position. In the short term, more dialogue between the developing countries and the rich nations is important.

4. Support for aboriginal peoples is still totally inadequate on a worldwide scale, and the independent organizations for aboriginal defence are often poorly funded and under-equipped. More finance, aid and media attention are needed to make prominent the plight of tribal people, and governments should be persuaded to use political pressure on their behalf.

5. Support for watershed protection. The loss of soil following deforestation is particularly severe when it occurs in a river catchment, and causes siltation and flooding. While a few countries have started watershed protection, these efforts are still often inadequate, and irreparable damage is occurring at an accelerating rate. In the absence of internal funds to support such measures, aid grants should be diverted to financing watershed replanting projects.

6. Research into forest resources is important if a logical and compelling case for forest protection is to be built up in the minds of citizens and governments. Screening plants for medicinal uses, research into new food sources, co-operation with tribal peoples to document their knowledge of plant uses, and collecting genetic material, are all vital stages in building up a more complete picture of forest resources. United Nations bodies, like the World Health Organization and the Food and Agriculture Organization, are already making a start, but far more effort on an international scale is necessary.

7. Education about trees and their importance. Education about the importance of trees is crucial, and will require innovative methods to put the message across; pictures, cartoons and puppet shows have all been used successfully on a small scale. Literacy programmes should concentrate on environmental issues when these are of importance to the students. Education should also extend to practical tree growing and maintenance, which could perhaps form part of standard school instruction, especially in rural areas.

8. Research into alternative planting strategies. Plantations will remain a necessary part of forest strategy for the foreseeable future, as will the use of exotic (i.e. introduced) species. However, the methods of afforestation need a radical re-examination in many parts of the world to avoid a repetition or continuation of the problems occurring from disease, wind blow and failure to produce good quality timber. In particular, the role of forestry needs to be examined with respect to other land use requirements, like food and fuel production, environmental conservation and amenity. The days of single purpose land use may well be numbered. The growing roles of forest farming or social forestry are a pointer to future developments, but will not succeed without help from research groups and funding agencies throughout the world.

9. Reducing the need for timber. Strategies for reducing timber requirements are essential, especially as more countries develop western-style expectations of consumption. Conservation techniques range from the use of more efficient woodstoves to techniques of paper recycling, along with more fundamental changes in attitude towards the use of resources, including re-assessment of throw-away products, the issue of over-packaging and so on. Some countries, like Sweden, have already gone a long way towards a national recycling strategy, with comprehensive legislation on paper recycling; others have hardly even made a start. Better international dissemination of information about the economics and technology of recycling would help to establish wider adoption of conservation programmes, while consumer pressure will be necessary to reduce the more wasteful uses of timber.

10. Reducing air pollution. The reduction in emissions of sulphur and nitrogen oxides is vital if forests in the rich countries are to recover and survive. International co-operation will be essential, because pollution can spread across borders to other countries and even other continents. Initiatives like the EEC pollution directive are to be welcomed in this respect.

These are only a few pointers of course; other actions necessary include extra control of the wildlife trade, more research into climatological effects and, most importantly, political will in the forest countries to tackle the underlying problems.

Sustainable forestry

In the longer term, a whole new attitude to forestry is needed, one where forests are managed in a much more holistic fashion, so that timber production does not automatically destroy the other possible functions of the forest as a food source, fuel supplier, environmental stabilizer and refuge for wildlife. Although aboriginal groups have practised holistic forest use for centuries, it has only recently been 're-invented' by modern foresters and farmers. In Australia, a system of 'permaculture' has been developed by Bill Morrison, where trees, crops and livestock are farmed on the same land, and through careful application of ecological knowledge, production is maximized without damaging the long-term fertility of the soil. In India, the concept of 'social forestry' has gained widespread support in the face of major resource and environmental problems brought on by deforestation. Social forestry aims to augment the dwindling forest by planting suitable tree species around villages, canals, railway lines and degraded forests to provide a mixture of resources and to make windbreaks and shelterbeds for farms. While the practice is certainly not perfect (the effects of reliance on exotic species and very high-density, intensive forestry are unclear), it is an important move in the right direction.

In the richer countries, public and private forestry is coming under increasing pressure to mix species, avoid straight lines of trees, include native species, landscape plantations and otherwise improve the stability and the amenity value of forests. There are signs that hedge removal is slowing down in Britain, and there is a growth of public support for forest preservation.

Unfortunately, these welcome moves are still small and widely dispersed; elsewhere forest removal is proceeding as quickly as ever, and planting is both unsympathetic to local conditions and badly planned. The urgent requirements outlined above are manifestly not being met at the moment, and the first priority of people interested in the future of forests is to see that this situation is reversed. As the publication, *Social Forestry in India* concludes: 'It is obvious that for the success of any programme of this nature, active involvement, co-operation and participation of the people from the various sections of the society is of the utmost importance.'*

* *Social Forestry in India*: Birla Institute of Scientific Research, New Delhi: Radiant Publishers 1984

10. Campaigning for forests

Environmental pressure groups

Environmental groups in the rich countries find themselves on the horns of a dilemma when trying to tackle the issues of world forests. The most pressing need is undoubtedly to stop the destruction of tropical moist forest, and there is considerable grassroots pressure to 'do something about it' within organizations like Friends of the Earth and Greenpeace, but there has been a lot of confusion about exactly what practical steps are possible. Northern groups are rightly reluctant to start moralizing about Third World problems when environmental conditions in the rich countries still leave so much to be desired. In any case, it is very difficult to see what can be achieved in the defence of forests thousands of miles away, even if interference seems justified. Tactless action will probably make matters worse rather than better, but on the other hand some developing countries are definitely sensitive to pressure from outside, especially on conservation issues.

A measure of the confusion about tackling tropical deforestation issues is illustrated by the doubts shown by Friends of the Earth and Greenpeace, probably the two most active environmental pressure groups in Britain. Friends of the Earth (FoE) have a good record of fighting habitat destruction in temperate areas, particularly in North America and Britain, and also in Australia, New Zealand and many European countries. The current Countryside Campaign in the UK has helped focus attention on the plight of Britain's own woodland areas and FoE in America have had a noticeable effect on government policies towards wilderness areas.

However, neither group has really got to grips with tropical forest destruction. Both Greenpeace and Friends of the Earth have talked about mounting a campaign for some years, and tropical forests were the special focus for the FoE international meeting in 1982. The discussions always flounder on exactly what to do.

One encouraging initiative is that FoE (UK) have recently sponsored the environmental research group Earth Resources Research to carry out work on the involvement of British firms in the tropical timber trade. FoE is hoping that a consumer campaign on these products will gain enough momentum to have a real effect on whether firms deal in tropical hardwoods or not, or will help persuade them to seek timber produced in a sustainable manner, following the successful boycott on whale products a few years ago which helped force manufacturers to find alternatives to the sperm whale oil used in their products. Even this approach is open to question, because any fall in timber demand will in the short term do damage to economies relying on timber, even if a great reduction in felling is needed in the future. It will be interesting to see how the campaign takes off. FoE and Greenpeace certainly defied their critics by making whales a world-famous issue. Tropical forests are already a powerful emotional focus and many people want to help 'preserve them' without really knowing how to start. It is not yet clear how this support can be channelled effectively.

Several groups have avoided the more difficult political issues by concentrating on fund-raising to sponsor specific reserves or education projects. Of these the largest by far is the World Wildlife Fund. The recently formed Earthlife Foundation may provide a valuable new pointer to how groups will develop in the future; it is raising cash specifically to buy a rainforest reserve in Cameroon.

One of the largest 'tropical rainforest' campaigns to date is being run by the World Wildlife Fund (WWF). The campaign follows a series of similar actions by WWF on other issues and has wide international backing. The World Wildlife Fund is a voluntary body, based in Geneva, with national groups throughout the world. It is closely linked with the International Union for the Conservation of Nature (IUCN), which provides the theoretical research and background to international conservation, while WWF co-ordinates the fund-raising and organizes field workers and projects.

Tropical forests represent something of a new departure for

WWF. Until recently it has concentrated almost exclusively on saving individual species of animals from extinction; raising money for colourful and popular animals like the panda (WWF's symbol) and large African mammals. Its principal role has been in purchasing reserves, although latterly it has been involved in extensive educational projects as well.

IUCN and WWF broadened their scope of activities significantly in the 1980s with the launch of the World Conservation Strategy, a document summarizing a number of environmental 'crisis areas' and outlining strategies and priorities for world conservation. The organization also called for member countries to produce their own 'national programmes' for implementing the strategy. About 40 countries, including Britain, have already done so.

Following the broader issues outlined in the strategy, and responding to pressure from WWF supporters at grassroots level, the organization launched the tropical forests campaign in 1983. Initially it was going to be a 'forests and primates' campaign, because WWF workers did not believe that forests alone would attract attention; in the event they have been proved wrong. The campaign aims to raise funds to buy forest reserves and to persuade national governments to adopt more forward-looking policies towards their forest areas. The initiative was widely praised, especially as an international voluntary group was effectively putting pressure on governments to act more responsibly towards the environment.

So far so good, at least on paper. However, behind the field workers and administrative staff, the organization is not really as 'independent' as it seems. WWF receives essential core funding from a group of about 1,000 people, who have each agreed to give a substantial sum each year, providing a base-line of about a million pounds even before public fund-raising and additional grants. While this comparative affluence has allowed the WWF to become the powerful organization it is today, it has also brought its own problems. Many of the funders are business people, and some represent a number of distinctly unpleasant corporations, including companies involved in mining, logging and other ecologically undesirable activities, often carried out with scant regard for the environment.

The links with business made little difference when the WWF

was simply raising money to save pandas. But once it started to look at the wider issues there were immediate conflicts of interest, with the people making the major donations a potential target for major criticism. In the event, the organization has ducked the issue. A glance at the World Conservation Strategy shows that behind all the fine ideals there are no concrete proposals at all for how most of the changes are going to be made, and particularly for how the major political issues, like control of transnational companies and implementation of land reform, are to be tackled in practice. Embattled politicians in the Third World are hardly going to respond to vague pleas about the genetic variability of the planet when they are up to their necks in debt to the IMF and trying to sort out a famine or an upsurge in violence. In many cases the strategy has obviously gone straight into the wastepaper bin, and the national plans that have been drawn up do not necessarily bear much resemblance to what is happening in practice.

This weakness is doubly unfortunate. Not only are the basic suggestions of the World Conservation Strategy essential, the WWF is probably the only body outside the United Nations with the funding or clout to make much impression in these areas. But if they are going to do so they will have to take a careful look at how the organization is funded and influenced. The many dedicated and competent WWF field workers and fund-raisers risk being let down by the structure of the organization as a whole.

Another smaller fund-raising group is Earthwatch, set up by a film-maker who made an award-winning film about the Cameroon rainforest, then set out to raise the funds to buy it as a nature reserve. This type of independent funding may well become more common in the future, and the possibilities for financing specific projects in countries offer at least a first-response method of forest protection. It would be fair to say that in the rich countries activism has definitely taken a second place to fund-raising. In the Third World, where the need is more pressing, there still appear to be relatively few grassroots movements for forest protection. One inspirational exception to this is the Chipko movement in northern India. Chipko (literally 'hugging the trees') developed because timber merchants were felling extensive forest tracts in India, leaving the local people with no fuel or fodder. In response to this, a group of women in one village directly challenged the timber extractors by putting their arms around trees which were to be felled, blocking the way with their own bodies. Not knowing how to respond to such actions, the timber fellers initially withdrew; when they returned they found the movement had spread throughout the area and involved people of all ages and both sexes with hundreds of people prepared to defend forests by non-violent means. Chipko has not just prevented localized felling. The issues raised in one area have spread further, promoting fierce debate about the role of forestry and, indirectly, helping to promote social forestry programmes.

Perhaps India is an ideal place to see the beginnings of such non-violent direct action, with its Gandhian tradition of 'Satagraha', peaceful protest usually involving a fast. Satagraha is Sanskrit and literally translates as 'insistence on truth', and the Chipko movement's main success has been in disseminating the truth about forest abuse. It is to be hoped that the lead shown by a few Indian villagers will be followed up more widely by peasants losing their resources through greedy and short-term forest exploitation.

11. Taking part

Forest abuse is neither inevitable nor unstoppable, but it will take a lot of effort to reverse current trends. I hope that by the time you finish this book, you will feel sufficiently concerned and angry to want to do something about it. There are a growing number of people already active on forest issues, working individually or (better) in groups, challenging short-term thinking and seeking more sustainable ways of using trees.

If you want to help, the first stage will probably be to contact organizations interested in forests. A representative list is given in the resources section (pp. 120–9). Working within an existing organization has a lot of advantages, but if you can't find one which is active in the areas you are concerned with, don't be afraid of starting your own; apart from anything else, the existence of a possible 'rival' group can sometimes spur the big groups into taking up an issue themselves. Small, local groups are often the most effective at fighting local compaigns such as saving a wood from felling.

The first main task is to educate ourselves and, in turn, those around us in workplaces, schools, colleges, clubs, unions, churches and environmental groups. Learn to look for environmental issues behind the headlines, read papers of different political persuasions and from different countries, and try to get a regular update on environmental and Third World issues.

Putting information across to other people can seem more daunting, but there are resources available to help, including tape slide sets, posters, leaflets and books. Forests are such an emotive issue that use of prepared material can often hold an audience's attention even if the speaker is not particularly experienced. At the same time, letters or articles in local papers, action within churches and local union branches, and working on local council members are all important. Using an official organization to add weight to a particular protest is effective and, as research continues and culprits are identified, a steady stream of protests may help persuade companies involved in bad forest exploitation to change their priorities. Politicians are also an important resource to be tapped; whatever their political allegiance, they may have some sympathy with the forest issue.

Writing about agribusiness, Susan George admonished 'study the rich and the powerful, not the poor and the powerless', making the pertinent point that earnest dissertations on peasant organizations are often useful to their oppressors to enable them to keep things the way they are. The same is true for forest issues. Most of the work needed is not in the jungles or the forests, but in tedious and painstaking research into the companies involved in forest abuse: who is being bought out; which organizations are included; what links there are between governments and forest industries and who stands to gain. This information can then be used by workers and activists to help bring about change.

In the final analysis, forest protection is not just about personal and cultural change. This change is only going to come from within people themselves, rather than being imposed by any government, so that any work you do should also involve a personal re-evaluation of your own lifestyle. This change does not have to be pretentious or holy. Actions like planting trees or reducing paper waste might seem like spitting into the wind in the face of environmental damage on the scale described in this book, but they are vital prerequisites to any real change.

Bibliography

Listing complete references for everything in this book would take up more space than is available in a short introduction to the subject. Instead, I will suggest a number of titles which expand on the information given here, and if there are specific facts for which readers would like a reference, or wish to query, I will be happy to answer any letters sent through the publisher.

Several books by the Worldwatch Institute and Earthscan are cited; this does not necessarily imply total agreement with everything they say, but both provide concise information on global issues. Worldwatch Papers are available from Third World Books, Stratford Road, Birmingham, UK, and Earthscan titles from Earthscan, 10 Percy Street, London W1P ODR.

Deforestation and its effects

Most of the up-to-date books on deforestation inevitably concentrate on the tropics. An excellent starting point are two by Norman Myers: *The Sinking Ark*, Oxford: Pergamon 1980 and *The Primary Source*, New York: W.W. Norton 1984. The first book looks at the general issue of disappearing species, and has a large section on tropical moist forest, while the second concentrates entirely on the tropical forest issue. For a shorter introduction, with additional emphasis on forest dwellers, Catherine Caufield's *Tropical Moist Forests* London: Earthscan 1983, is probably the best buy.

Detailed information on forest peoples is contained in a number of reports from the Minority Rights Group, especially Hugh O'Shaughnessy and Stephen Cory, *What Future For The Amerindians of South America?*, London: MRG 1977. The role of the USA, and the resistance by grassroots members of the catholic church in Latin America, is movingly described by Penny Lernoux in *Cry For The People*, London: Penguin 1982. The group Survival International also publish good information on forest dwellers, and for an account of the effects of civilization on aboriginal tribes in Papua New Guinea try *Oh What A Blow That Phantom Gave Me!* by Edmund Carpenter, St Albans: Paladin 1976.

Disappearing species are covered in dozens of natural history books, but for a slightly different approach try the short book by Erik Eckholm, *Disappearing Species: The Social Challenge*, Worldwatch Report, 1979 and two books by Robert and Christine Prescott Allen, *What Use Is Wildlife?* and *Genes From The Wild*, Earthscan,1984, which describe the use of wild plants and animals for crop breeding. Soil erosion problems are dealt with in *The Other Energy Crisis: Firewood* by Erik Eckholm, Worldwatch 1975, and by Norman Myers and Catherine Caufield in their publications. For a readable guide through the intricate arguments of the meteorologists, try John Gribbin, *Carbon Dioxide, Climate and Man*, Earthscan 1980. *Desertification* by Alan Grainger, Earthscan 1982, introduces the whole subject of desert formation, and the current spread of deserts.

Deforestation and afforestation in temperate forests

The general literature about afforestation is not as comprehensive as that on deforestation. Two good books about Britain's position are *The Theft Of The Countryside* by Marion Shoard, London: M.T.Smith 1980, which attacks the farming industry for its role in rural abuse and provoked a storm that made Ms Shoard cordially hated by most farmers in the UK, and *Crisis And Conservation* by Chris Rose and Charlie Pye-Smith, London: Penguin 1984, which includes a good introductory chapter on forests. More detailed information can be found in *The Future Of Forestry* by Richard Grove, Stanfont-St John: British Association of Nature Conservationists 1983, and *The Loss Of Wildwoods* by Grove and Chris Rose, London: Friends of the Earth 1984. The Ramblers Association have produced a couple of good papers on afforestation, in response to the Centre for Agricultural Strategy paper, 'Strategy For The UK Forest Industry', CAS 1980, which is interesting in that it shows what foresters were thinking about a few years ago.

The journal *Ecos*, published by the British Association of Nature Conservationists, is a must for anyone interested in keeping up to date with events in British forests.

Tree diseases

Tree diseases are extensively covered by the specialist press but not as well for the general reader. Robert Lamb's book *World Without Trees*, London: Wildwood 1979, is a general account of forest problems but includes a great deal on diseases, and especially Dutch Elm Disease, and Gerald Wilkinson's *Epitaph For The Elm*, London: Hutchinson 1980, which is devoted entirely to DED, is a pictorial rather than academic treatise. Quite an old book which is still worth looking out for is *The Advance Of The Fungi* by E.C.Large, London: Cape 1940. Forestry Commission publications give details on actual diseases in Britain. The spraying issue is tackled classically by Rachel Carson in *Silent Spring*, London: Penguin 1966, and more recently by Robert van den Bosch in *The Pesticides Conspiracy*, New York: Doubleday 1978. The leaflet 'Spray Drift' from the Soil Association pinpoints the inevitable hazards involved with spraying.

Pollution damage

There has been a vast number of reports and proceedings about acid rain and forests over the past five years. A good general introduction is Steve Elsworth's *Acid Rain*, London: Pluto 1984, and a more detailed analysis of ecological effects can be found in *The Acid Rain Controversy* by Mark Barrett and Nigel Dudley, London: Earth Resources Research 1984. Other useful publications include the Nature Conservancy Council's report *Acid Deposition And Its Implications For Nature Conservation In Britain* by Garry Fry and Arnold Cooke, London: NCC 1984, which is more academic but is the first really to look at wildlife implications, and *Air Pollution, Acid Rain And The Future Of Forests* by Sandra Postel, Worldwatch 1984, which gives a US perspective on forest damage.

Where it is happening

Reports of forest damage are scattered and fragmentary. Two overviews that have been useful are the report on *Conversion Of Tropical Moist Forests*, by Norman Myers, National Academy of Sciences 1980, and *Tropical Forest Resources* by J.P. Lanly, Food and Agricultural Organization 1984, both of which attempt to give an overview of the state of tropical forests. The famous *Global 2000 Report To The President*, London: Penguin 1980, which was commissioned by Jimmy Carter and completely ignored by Ronald Reagan, has a good analysis of the likely changes to forests in the next few decades. Norman Myers's two books referred to earlier give brief regional surveys.

Continuing information updates are more difficult to come by. Various magazines cover deforestation, including *The Ecologist*, the *IUCN News* and *Deforestation And Development* from the European Environmental Bureau. The World Wildlife Fund is producing more information on forests and the magazine *New Scientist* has carried some useful articles. Friends of the Earth (UK) have produced a first issue of *TRF News*, which may be a regular event if enough people subscribe. Various other national FoE groups cover TRF, including one in Malaya which publishes regular newsletters. There is still a definite need for a regular worldwide publication on forest abuse, accessible to the non-specialist.

Population

Generations of environmentalists have been brought up on Paul Ehrlich's *The Population Bomb*, which whatever its merits in drawing attention to population problems has spread a desperately simplistic philosophy of population. (As I have referred to so many Worldwatch publications I should say that some of their population titles perpetrate much the same fallacy.) Susan George demolishes the overpopulation thesis in her analysis of food shortages in *How The Other Half Dies*, London: Penguin 1979, which is an excellent starting point for anyone interested in development issues. Kathleen Newland's short paper 'Women and population growth: choice beyond childbearing', Worldwatch 1977, is a welcome exception to what I said above about Worldwatch. She analyzes the links between education and emancipation of women and slowing of population growth.

Land reform

Sven Lindqvist has done a moving survey of land in South America, which is now slightly out of date but still worth buying: *Land And Power In South America* London: Penguin 1979, 1983 while Clare Whitmore gives a readable introduction to the issues in *Land For People: Land Tenure And The Very Poor*, Oxford: Oxfam, and Kathleen Newland discusses people's movements in *International Migration: The Search For Work*, Worldwatch 1980.

The multinationals

Multinational companies have been given a fair mauling by the food lobby and readers could start here with Susan George's book mentioned above, and also *Food First* by Frances Moore Lappe and Joseph Collins, London: Abacus 1978. For a more detailed survey of a specific area, try *Agribusiness In Africa* by Barbara Dinham and Colin Hines, London: Earth Resources Research 1983. For regular informatio see *Multinational Monitor*, a journal published in the USA, and *New Internationalist* in Britain. There are also fairly regular publications on specific issues like Rio Tinto Zinc and the coffee trade, and Counter Information Services (UK) has done some reports on transnational companies as well as providing information on international banking and the movement of finance. Anthony Sampson has produced a number of books on multinational companies and banking: *The Sovereign State*, London: Coronet 1977 is about ITT.

Capitalism

Two good general introductions to the role of capitalism are *Due South* by Jeremy Hill and Hilary Scannell, London: Pluto 1983 and *Hard Times* by Bob Sutcliffe, London: Pluto 1983. *The Creation Of World Poverty* by Teresa Hayter, London: Pluto 1981, provides an alternative view of the issues raised in the Brandt report *North South* and challenges the role of the rich countries in under-developing the Third World. For a voice from within the south, Franz Fanon is still a powerful advocate, especially in his book *The Wretched Of The Earth*, London: Penguin 1970. Several specific cases of multinational activities have been written up by War on Want, see especially *Tomorrow's Epidemic?* by Mike Mueller, London: War on Want, 1978, on the tobacco trade in developing countries.

Forests for the future

Myers's book *The Primary Source* includes some more hopeful visions of the future. A short report giving a useful overview of some recent developments is *Planting For The Future* by Erik Eckholm, Worldwatch 1980, while some of the ideas of sustainable forest use are summarized in *Permaculture One And Two* by Bill Morrison, Thorsons, and *Forest Farming* by Robert Hart Davis, London: Watkins 1976. Richard St Barbe Baker's last book *My Life My Trees*, Thorsons, is visionary in its respect to trees (and prehistoric in its politics).

For anyone actually wanting to get their hands dirty and plant trees or manage woodlands, two very clear and comprehensive handbooks are available from the British Trust for Conservation Volunteers (London): *Tree Nurseries* which is a short introduction to growing trees from seed, and *Woodlands* which is a much larger volume dealing with all aspects of forest management from a nature conservation perspective.

Politics of change

The writings of E.F. Schumacher and Ivan Illich will probably already be familiar to most people; *Small Is Beautiful*, London: Abacus 1975 by Schumacher and *Tools For Conviviality*, London: Fontana 1975 by Illich are both controversial starting points. Books written from a left perspective include *Ecology and Politics* by Andre Gorz, London: Pluto 1980, and *Socialism and Ecology*, a pamphlet by Raymond Williams, London: Socialist Environment and Resources Association 1982. Also try *Fighting For Hope*,

London: Chatto 1984 by Petra Kelly, one of the leading figures in the German Green Party and *From Red To Green*, London: Verso 1983, conversations with Rudolf Bahro covering his doubts about socialism and his joining of the Green Party. For general information about how to campaign, try *What Do We Do After We've Shown The War Game?* by Daniel Pletsch, London: CND 1982, which provides good advice for pressure group work, *The A-Z Of Campaigning* by Des Wilson, London: Heinemann 1984, and D.McShane, *Using The Media*, London: Pluto Press 1979.

Useful addresses

This is by no means a complete list of organizations involved in forests. It attempts to include a selection of the main groups involved in campaigning for and acquiring natural British forests, and also those interested in forests or forest-related issues abroad.

NON-GOVERNMENTAL ORGANIZATIONS INVOLVED WITH FORESTS

ORGANIZATIONS MAINLY ACTIVE WITHIN BRITAIN

British Association of Nature Conservationists,

Rectory Farm, Stanton-St John, Oxfordshire OX9 1HF; telephone 0867 735 214

BANC is a relatively new, young and radical conservation group whose main function to date has been the production of *Ecos*, a quarterly review of conservation which has already had an important effect on attitudes to British conservation. Also produces occasional reports and seminars.

Council for the Protection of Rural England,

4 Hobart Place, London SW1W OHY; telephone 01 235 9481

Council for the Protection of Rural Wales/Cymdeithas Diogelu Cymru Wledig,

High Street, Welshpool, Powys SY21 7JP; telephone 0938 2525

Both CPRE and CPRW exist to campaign for better controls over agriculture and forestry, initially concentrating on the setting up of national parks but now more broadly based and more radical than the middle-class and middle-aged membership would suggest.

Friends of the Earth,

377 City Road, London EC1V 1NA; telephone 01 837 0731

FoE is an international organization with branches in about 40 countries, including some possessing tropical forests. In Britain FoE has about 250 local groups, some employing full time staff and others being entirely voluntary. FoE have been active in opposing habitat destruction for over a decade and have recently stepped up their activities at home with a well-publicized Countryside Campaign. They are also developing a tropical forests campaign based on British trade in timber products. Membership at local and national levels; FoE will send the address of your nearest local group.

Greenpeace,

36 Graham Street, London N1 8JX; telephone 01 251 3020

Another international organization. UK groups are confined to fund-raising, ostensibly because Greenpeace does not want to get the blame for local independent groups doing dangerous stunts (sometimes FoE UK must wish they had a similar arrangement). Greenpeace is predominantly an ocean orientated campaign group with well-publicized actions in defence of seals and whales and against sea dumping of nuclear waste. A recent concerted European chimney climb to protest against acid rain may mark the start of more land-based action and Greenpeace has definitely stated an interest in doing work on tropical forests at some time in the future.

Men of the Trees,

Turner's Hill Road, Crawley Down, Crawley, W. Sussex RH10 4HL; telephone 0342 712 536

The Men was started by Richard St Barbe Baker, one of the early visionaries of trees and the need for forests. They hold meetings and publish a regular journal but do not seem to be particularly involved in the nitty gritty of either conservation or tree planting any longer.

National Trust,

42 Queen Anne's Gate, London SW1H 9AS; telephone 01 222 9251

The NT exists to preserve both architectural and environmental heritage and is one of the largest landowners in the country, controlling about 1 per cent of the land area. It is also undoubtedly the biggest environmental organization, with about a million members. Although it has preserved important areas of woodland, it avoids any hint of pressure group activity and was severely criticized in 1983 for allowing the army access to some of its 'protected' land for training purposes.

Ramblers Association,

1–5 Wandsworth Road, London SW8 2LJ; telephone 01 582 6826

Started in the wake of the (largely communist-led) trespasses, to gain access to private land for hikers from the urban areas of northern Britain. The Ramblers are important to forestry because they have been the most consistent critics of modern British plantation methods and have undoubtedly been influential in promoting the changes that are now being considered.

Royal Forestry Society of England, Wales and Northern Ireland,

102 High Street, Tring, Herts. HP23 4AH; telephone 044 228 2028

Royal Scottish Forestry Society,

1 Rothesay Terrace, Edinburgh EH3 7UP; telephone 031 225 5561

These are both useful organizations especially for professional foresters, offering career advice, visits to forest estates, etc; also have local branches. Both societies have fairly conventional approaches to forest policy and conservation.

Royal Society for Nature Conservation,

The Green, Nettleham, Lincs. LN2 2NR; telephone 0522 52326

The RSNC is the co-ordinator of the 40 or so county naturalists' trusts, which are supported by voluntary membership, and have a limited number of full time staff. The bulk of their effort is put into education and the acquisition and management of reserves, although they will also campaign on specific nature conservation issues and have become slightly more radical recently in criticizing the government over their Wildlife and the Countryside Act, and its implementation. Still fairly conservative.

Royal Society for the Protection of Birds,

The Lodge, Sandy, Beds. SG19 2DL; telephone 0767 80551

The largest nature conservation group in Britain with about 300,000 members. Although interested specifically in birds, the RSPB manages 80 reserves, many of which include areas of woodland. Important in that it brings many people into contact with conservation ideas who would otherwise not do so. Starting to address slightly wider issues through the pages of its magazine *Birds*.

Tree Council,

35 Belgrave Square, London, SW1X 8QN; telephone 01 235 8854

The Tree Council is an educational charity aimed at promoting tree planting in Britain, mainly for environmental and aesthetic reasons. They have limited funding for selected projects.

Woodland Trust,

34 Westgate, Grantham, Lincs. NG31 6LL; telephone 0476 74297

Aims to protect natural British woodland, both by planting trees and by buying reserves. Now also involved in community woodland. The Woodland Trust is still quite a young organization and has had some notable successes in preserving ancient and ecologically important woodland; not noticeably interested in the wider forestry issues however.

ORGANIZATIONS MAINLY ACTIVE ABROAD

(Both Friends of the Earth and Greenpeace, listed above, also have an international perspective.)

Christian Aid,

PO Box 1, London SW9 8BH

Christian Aid works on development projects in the Third World. Like Oxfam (see below) they are gradually changing from a purely crisis-relief function to sponsoring more long-term projects, including education about ecology and forests. Co-sponsor of *New Internationalist* magazine.

Green Deserts,

Geoff's House, Rougham, Bury St Edmunds, Suffolk; telephone 0359 70265

Interested in re-afforestation in the desert regions, both practically with a project in the Sudan and theoretically through education work and running the Rougham Tree Fair. Engaged in some interesting experimental work, including the use of puppet shows to teach the facts about forest loss and desertification to people without reading skills.

Intermediate Technology Development Group and Appropriate Technology for Forestry,

9 King Street, London WC2E 8HN; telephone 01 836 9434

The Intermediate Technology Development Group was started by E.F. Schumacher (of *Small is Beautiful* fame) to help promote and develop intermediate technology in Third World countries. The group is now also working on small-scale developments in the north. Of particular interest here is the project to promote cheap and accessible wood-burning stoves to stretch resources of timber. The Appropriate Technology for Forestry is an offshoot of ITDG, providing a comprehensive source of expertise to assist developing countries in all aspects of foreign management.

International Tree Crops Institute,

2 Convent Lane, Bocking, Braintree, Essex CM7 6RN

ITCI was set up to promote the use of trees as foods, develop the use of natural tree products and disseminate information to people working on forests, especially in the Third World. A related organization also exists in the USA, and both publish a regular journal.

Oxfam,

274 Banbury Road, Oxford OX2 7DZ; telephone 0865 56777

As well as famine relief, Oxfam now includes issues like transnational involvement in trade and development, the drugs trade and ecological issues in its brief. Co-sponsor of *New Internationalist* (with Christian Aid) which is one of the best general development journals likely to be read by the average non-political person interested in the issues. Some good publications.

War on Want,

467 Caledonian Road, London N7 9BE; telephone 01 609 0211

The most political of the relief organizations; its tradition of backing causes it believes in has both won and lost it support. (Many people have never forgiven WOW for providing funds for the Grunwick strikers.) Basically still a relief organization but also produces occasional reports. Consistent critic of transnationals.

World Wildlife Fund (UK),

Panda House, 12-13 Ockford Road, Godalming, Surrey
GU7 1QU; telephone 04868 20551

Despite the criticisms raised in the main text of this book,
WWF is still the most influential voluntary organization for
promoting and funding forest conservation on a worldwide
scale, and the British organization has survived recent criticisms
relatively unscathed. Principal aims are fund-raising, especially
for the International Union for the Conservation of Nature, but
it also funds work by other groups.

OFFICIAL ORGANIZATIONS INVOLVED WITH FORESTS

ORGANIZATIONS ACTING WITHIN BRITAIN

Arboricultural Association,

Ampfield House, Ampfield, Romsey, Hampshire SO5 9PA;
telephone 0794 68717

Professional association for foresters, list of consultants
available, also publications on tree care etc.

Association of Professional Foresters,

Brokerswood Park, Brokerswood, Westbury, Wilts. BA13 4EH;
telephone 0373 822 238

Professional association for individuals and corporations
involved in private forestry, strong lobby with the government;
not overpoweringly interested in conservation and development
issues.

Commonwealth Forestry Institute,

University of Oxford, South Parks Road, Oxford, OX1 3RB;
telephone 0865 50156

Houses the Commmonwealth Forestry Association and the
Commonwealth Forestry Bureau. The former is a very broad
professional association ranging from timber merchants to

conservationists, while the latter acts as a data base and publishes news, abstracts etc. Influential especially in tropical forestry.

Countryside Commission,

John Dower House, Crescent Place, Cheltenham, Glos. GL50 3RA; telephone 0242 21381

Countryside Commission for Scotland,

Battleby, Redgorton, Perth PH1 3EW; telephone 0738 27921

Government organizations, initially set up to help manage the national parks, now a more general amenity function including grants for footpaths, helping in formation of country parks, Areas of Outstanding Natural Beauty (AONBs) and Heritage Coasts. Often criticized as being weak in standing up to large forestry and farming interests, organizer of large uplands survey and strategy including much work on forestry. Disburses grants for planting trees.

Forestry Commission,

231 Corstorphine Road, Edinburgh EH12 7AT; telephone 031 334 0303

Government forestry body, owns large areas of forestry plantation and ancient and semi-natural woodland, also has some policy control over private ownership. In the past the FC has acted insensitively towards both planting strategies and acquisition of land (partly because conservation was not part of their brief), now severely cut back by the Tory government and forced to sell off large forest areas. Have changed their views at least officially, but still a long way to go in practice and much of the local forestry impact (i.e. planting patterns, amount of hedge clearance etc.) are in control of local managers, so performance differs around the country.

Nature Conservancy Council,

19–20 Belgrave Square, London SW1X 8PY; telephone 01 235 3241; also many regional offices

The NCC is charged with nature conservation in Britain, administering local and national nature reserves, commenting on afforestation schemes etc. Unfortunately, the NCC has remained relatively toothless, being headed largely by civil servants and latterly people nominated by the Country Landowners Association and the National Farmers' Union. When the last director, Sir Ralph Verney, actually followed through his statutory obligations in declaring a Somerset site as being protected, the local Conservative MP told him he would lose his job, which indeed he did. For a succinct account of why the NCC cannot fulfil its obligations until the law is changed see *Implementing the Act* by W.M. Adams, Stanton-St John: British Association of Nature Conservationists 1984. Local NCC officials often very helpful.

Timber Growers United Kingdom Limited,

Agriculture House, Knightsbridge, London SW1X 7NJ; telephone 01 235 2925

Timber Growers Scotland Limited,

6 Chester Street, Edinburgh EH3 7RA; telephone 031 226 3475

Both lobby the government and Forestry Commission on behalf of private forestry.

ORGANIZATIONS WORKING MAINLY ABROAD

Commonwealth Forestry Institute,

Listed above, interests both within the UK and overseas.

Food and Agriculture Organization (FAO),

Via delle Terme di Caracalla, 00100, Rome, Italy

The part of the United Nations most concerned with forestry. Until recently its interest has been almost wholly in management and extraction rather than conservation. There are signs that this attitude is changing, although still not quickly enough.

At the time of going to press, Friends of the Earth is launching a campaign with the following objectives:

1. Persuading the tropical hardwood timber trade in consumer countries to adopt a code of conduct stressing the need for sustainable tropical forestry operations, replanting of felled areas and protection of the tropical forest eco-system. Friends of the Earth will be running a consumer pressure campaign in developed countries like the United Kingdom, the United States of America, Japan and other European countries to help bring this about.

2. Changing the Government Aid programmes of the United Kingdom and other European countries to promote the sustainable use of tropical forests and to withhold aid from capital projects which damage the tropical moist forest eco-system.

3. Raising money in developed countries to help Friends of the Earth groups in tropical forest producer countries to run campaigns promoting Forestry Conservation Strategies, which will encourage sustainable forestry and replanting schemes.

4. Conducting future research into the environmental, economic and social benefits of preserving tropical rain forest eco-systems.

5. Launching a public awareness campaign of the global and regional problems caused by tropical deforestation, including adverse climatic changes, the wholesale extinction of species, and the loss of vital natural resources.

Index